THE INTERNATIONAL SERIES OF
MONOGRAPHS ON CHEMISTRY

EDITOR: J. S. ROWLINSON

VIBRATIONAL AND ROTATIONAL RELAXATION IN GASES

BY

J. D. LAMBERT

Fellow of Trinity College, Oxford

CLARENDON PRESS · OXFORD · 1977

Oxford University Press, Walton Street, Oxford OX2 6DP

OXFORD LONDON GLASGOW NEW YORK
TORONTO MELBOURNE WELLINGTON CAPE TOWN
IBADAN NAIROBI DAR ES SALAAM LUSAKA ADDIS ABABA
KUALA LUMPUR SINGAPORE JAKARTA HONG KONG TOKYO
DELHI BOMBAY CALCUTTA MADRAS KARACHI

British Library Cataloguing in Publication Data

Lambert, J D
Vibrational and rotational relaxation in gases.
– (The international series of monographs on chemistry).
1. Energy transfer 2. Gases 3. Molecules
I. Title II. Series
533′.7 QC175.16.M6 77–30262

ISBN 0-19-855605-5

Typeset by Hope Services, Wantage
Printed in Great Britain
by Morrison & Gibb Ltd., Edinburgh

TO

BERTRAM LAMBERT, CYRIL HINSHELWOOD

AND

ARNOLD EUCKEN

IN GRATEFUL MEMORY

PREFACE

This book is concerned with the interchange of translational, vibrational, and rotational energy between molecules in collision. Experimental studies in this field started some forty years ago with acoustic measurements on ultrasonic relaxation, and the first textbook on energy transfer, *Absorption and dispersion of ultrasonic waves* by Herzfeld and Litovitz, appeared in 1959. Shortly before this date shock-wave techniques had been introduced enabling relaxation measurements to be extended to higher temperatures and to a larger range of molecules. In 1961 Cottrell and McCoubrey's book *Molecular energy transfer in gases* appeared, giving a comprehensive survey of experimental results obtained by both methods. Both of these books were concerned almost exclusively with vibration-translation (V-T) and rotation-translation (R-T) transfers, and with the underlying semiclassical and quantum mechanical theories. These theories predicted that vibration-vibration(V-V) transfer must play an equally important role in vibrational relaxation. Experimental investigations on V-V transfer, obtained by acoustic and flash-photolytic techniques, were first presented at a Faraday Society Discussion on inelastic collisions in 1962. Advances were soon made by both techniques, and a Chemical Society Quarterly Review on V-V transfer by Lambert appeared in 1967. Stevens' book *Collisional activation in gases* appeared in the same year, followed a little later by a multi-volume treatise, *Transfer and storage of energy by molecules*, edited by Burnett and North. These review the state of the subject at that time. In the past decade entirely new techniques–laser-excited fluorescence, infrared chemiluminescence, and molecular beams–have been introduced, leading to very rapid development over the whole field. There is now available a large mass of experimental information about V-T and V-V transfer, also about vibration-rotation (V-R) transfer, which has been shown to play an important role. Some interesting new results on R-T transfer have also appeared in the past few years.

The object of the present book is to give a concise survey of the entire field of modern experimental knowledge about energy transfer. It includes sufficient information about the experimental techniques to enable an assessment of their relative accuracy and limitations, and sufficient theoretical background to form a basic pattern into which the experimental data may be fitted. It assumes a normal undergraduate background in chemical physics, and is intended for advanced undergraduate study and for research workers and technologists who need reliable information about energy transfer phenomena. Some of the basic mathematical bookwork underlying the classical experimental methods, which is very adequately treated in standard works, is omitted.

For vibrational transitions, where relatively large energy quanta are involved, quantum-mechanical theory is generally applicable, and the theoretical treatment in the book is based almost entirely on the Schwartz, Slawsky, and Herzfeld (SSH) quantum-mechanical theory and its modifications. This is still the only theory widely used for quantitative comparison with experimental results. The relation of SSH theory to other and more exact theories, classical, semi-classical, and quantum-mechanical has been fully discussed in review articles by Takayanagi in 1965, by Rapp and Kassal in 1968, and by Secrest in 1973. For rotational transitions, where much smaller energy quanta are involved, quantum-mechanical theory may be applied to hydrogen and hydrides (where the spacing of rotational levels remains relatively large), but classical theory is more appropriate for heavier molecules. Both types of theory are therefore briefly discussed and used in the interpretation of the experimental data. The elementary discussion of molecular collisions in Chapter 1 extends only far enough to explain the concepts used in the book, and is not intended to be complete. A full discussion of molecular collision processes, both elastic and inelastic, will be found in M. S. Child's book, *Molecular collision theory* (Academic Press, 1974).

Oxford J.D.L.
May 1976

ACKNOWLEDGEMENTS

I am grateful to Professor J. S. Rowlinson for kindly reading through the completed work in typescript and giving me his comments, to Dr. A. B. Callear, Dr. J. H. W. Cramp, and Dr. C. J. S. M. Simpson for supplying photographs, and to the Royal Society, the Chemical Society, the Institute of Physics, the North-Holland Publishing Company, and John Wiley and Sons Inc. for permission to use copyright figures and tables.

CONTENTS

UNITS AND EQUATIONS

The efficiency of an energy transfer process is expressed throughout this book in terms of the *collision number*, Z_{10} $(=1/P_{10})$, as defined in § 1.4.2. Data published on other units have been converted to collision numbers by the procedures described. Extensive tables of collision numbers calculated from experimental data published before 1965 are given in B. Stevens (1967), *Collisional activation in gases*, Pergamon, Oxford, and these were used where available. Data for V-V transfer published before 1973 were taken from Bradley Moore's review (1973), Vibration-vibration energy transfer. *Adv. chem. Phys.* **23**, 41, Table 2.

It should be noted that energy transfer processes are not chemical reactions, and the thermodynamic sign conventions used for heats of reaction etc. do not apply. The symbol $+\Delta E$ denotes an amount of internal energy transferred to translation. Thus,

$$SO_2^*(\nu_1) + SO_2(0) \rightarrow SO_2^*(2\nu_2) + SO_2(0) + 110\ \text{cm}^{-1}$$

or

$$SO_2^*(\nu_1) + SO_2(0) \rightarrow SO_2^*(2\nu_2) + SO_2(0) + \Delta E$$

represents an *exothermic* process, as may be seen on the energy-level diagram, Fig. 4.6, while

$$SO_2^*(\nu_1) + SO_2(0) \rightarrow SO_2(0) + SO_2^*(\nu_3) - 210\ \text{cm}^{-1}$$

or

$$SO_2^*(\nu_1) + SO_2(0) \rightarrow SO_2(0) + SO_2^*(\nu_3) - \Delta E$$

represents an *endothermic* process, where internal energy is increased at the expense of translational energy. Each of the symbols in the equations represents an *energy level*, and the two sides of the equation balance numerically.

Numerical units throughout the book are cgs and esu, which are used almost universally in the literature of the subject. Energies and pressures may be converted to SI units by the relations:

$$1\ \text{cal mol}^{-1} = 0{\cdot}35\ \text{cm}^{-1} = 4{\cdot}184\ \text{J mol}^{-1}$$

$$1\ \text{cm}^{-1} = 29{\cdot}98\ \text{GHz}$$

$$1\ \text{atm} = 760\ \text{torr} = 101{\cdot}325\ \text{kN m}^{-2}$$

The numbering of vibrational fundamentals of polyatomic molecules, ν_1, ν_2, ν_3 etc., follows the conventions of Herzberg.[89] For triatomic molecules energy levels are expressed in the form (v_1, v_2, v_3). For linear triatomic molecules, such as CO_2, where ν_2 is doubly degenerate, (v_1, v_2^l, v_3) denotes a state with v_1 quanta of mode ν_1 excited, v_2 quanta of mode ν_2 with vibrational angular momentum 1, and ν_3 quanta of mode v_3.

SYMBOLS AND NOTATION

(Symbols not listed here are defined where they appear)

A	spectroscopic rotation constant (Table 3.9)
(A^2)	vibrational amplitude coefficient (Table 3.1)
A_{10}	Einstein coefficient (radiation) (2.13)
a	range of intermolecular forces (§ 1.3.1)
B	spectroscopic rotation constant (Table 3.9)
B	second virial coefficient (2.7)
B_{01}, B_{10}	Einstein coefficients (radiation) (2.13)
b	impact parameter (Fig. 1.2)
C	spectroscopic rotation constant (Table 3.9)
C_p	molar heat capacity at constant pressure
C_V	molar heat capacity at constant volume
C_1, C_2	heat capacity contributions of fundamental modes, 1, 2
C_s	total *vibrational* heat capacity (Ger. *Schwingungswärme*)
$C_{\mathrm{I}}, C_{\mathrm{II}}$	exponential factors in Lambert-Salter eqns (3.12)
C	exponential factor in Polanyi-Woodall eqn (3.17)
c	speed of light
D	diffusion coefficient (2.16)
D	Debye unit (dipole moment) (10^{-18} esu cm)
d, d_0 etc.	distances, defined when used
$+\Delta E$	quantity of internal energy transferred to translation
$+\Delta E_{\mathrm{vib}}$	quantity of vibrational energy transferred to translation
$+\Delta E_{\mathrm{rot}}$	quantity of rotational energy transferred to translation
f	acoustic frequency (Hz)
ΔH	enthalpy of chemical reaction
h	Planck's constant
$\hbar$	$h/2\pi$
$I, I_{\mathrm{A,B,C}}$	moments of inertia (Table 3.9)
$I(E_j)$	'surprisal' corresponding to state distribution $P(E_j)$ (§ 3.6)
J	rotational quantum number

K	rotational quantum number
k	thermal conductivity (2.14)
k	Boltzmann constant
k	rate constant (1.7)
M	relative molar mass ('molecular weight')
M_I	Mach number
m	mass of a molecule
n	number of molecules in unit volume (number density)
P_{10}	probability per collision of 1-0 transition (1.4)
$P_{k-l(b)}^{i-j(a)}$	probability of simultaneous transitions, *i-j* and *k-l* (3.2)
P_0	steric factor in SSH theory (3.2)
$P(E_j)$	actual state distribution (§ 3.6)
$P^0(E_j)$	*a priori* state distribution (§ 3.6)
p	pressure
Q	collision cross-section for energy transfer (1.8)
R	gas constant
r	intermolecular separation
r_c	separation at classical turning-point (Fig. 1.1)
r_m	separation at potential minimum (Fig. 1.1)
T	temperature (K)
T_{TR}	translational temperature
T_{VIB}	vibrational temperature
T_{ROT}	rotational temperature
t_ν	period of an oscillation (§ 1.3.1)
$U(\nu)$	energy density of radiation field at frequency ν (2.13)
ΔU	energy of dimerization (dipole interaction) (Table 1.2)
V	molar volume
$V(r)$	intermolecular potential at separation, r (1.1)
$V^{i\text{-}j}$	'vibration' factor for *i-j* transition (3.2), (3.3)
$\mathbf{v}$	velocity
v	molecular speed
v^*	optimum molecular speed for transition (Fig. 1.6)
v	vibrational quantum number
W	speed of sound (2.2)
W_s	speed of a shock-front (§ 2.3.1)

x	mole fraction in a mixture (3.15)
Z	gas kinetic collision rate (1.2)
Z^*	'hard-sphere' collision rate (§ 1.2.3)
Z_{10}	collision number for a 1-0 transition ($=1/P_{10}$) (1.5)
Z_R	collision number for rotational relaxation
Z_{AB}	collision number for heteromolecular process, A + B
α	exponential parameter in repulsive potential (§ 1.4.1)
α^*	value of α for optimum velocity of approach (§ 3.1.2)
β	isothermal relaxation time ($= \tau_{VT} = \tau_{pT}$) (§ 2.2.1)
β_1, β_2	relaxation times for V-T transfer from fundamental modes, 1, 2
β_{12}	relaxation time for V-V transfer between modes 1 and 2
β_{AB}	relaxation time for V-V transfer between molecules A and B
β_R	rotational relaxation time
γ	ratio of specific heats, C_p/C_V
γ	frequency term in SSH vibrational factor (3.5)
ϵ	depth of intermolecular potential well (Fig. 1.1)
ζ	energy term in SSH theory (3.3)
η	shear viscosity
λ	acoustic wavelength
μ	reduced mass for a collision (§ 1.2.3)
μ	acoustic absorption coefficient (2.6)
μ	dipole moment (Table 3.3
ν	vibration frequency (wavenumber)
ν_{min}	frequency of lowest fundamental mode of polyatomic molecule
ν_1, ν_2	frequencies of modes, 1, 2 (Herzberg[89])
ξ	adiabaticity parameter (§ 1.3.1)
ρ	density of gas
σ	intermolecular separation at zero energy (Fig. 1.1)
σ^*	'hard-sphere' collision diameter (§ 1.2.3)
$\sigma_{i\text{-}j}$	total cross-section for *i-j* energy transfer (§ 2.8)

τ	radiative lifetime (§ 2.4.1)
τ_c	duration of a collision (§ 2.2.1)
τ_{VT}, τ_{pT}	isothermal relaxation time (§ 2.2.1)
τ_{pS}	adiabatic constant-pressure relaxation time (§ 2.2.1)
τ_{VS}	adiabatic constant-volume relaxation time (§ 2.2.1)
χ	displacement of an oscillator (Fig. 1.4)
ω	cyclic frequency ($=2\pi f$)

1

MOLECULAR COLLISIONS

1.1. Introduction

Molecules usually gain and lose vibrational and rotational energy in collisions. Polar molecules *can* be activated or deactivated by absorption or emission of radiation, but the radiative lifetime (which varies with ν^{-3}), is so long for infrared frequencies that this mechanism may be neglected under most conditions (see § 2.4). The nature of the collision process is thus of fundamental importance, and this chapter will consider aspects of the theory of collisions between real molecules which are important for energy transfer. We shall first discuss the model used for explaining collisions between inert gas molecules, which have no internal degrees of freedom, and then apply the same principles to vibrating and rotating diatomic molecules, and finally to polar molecules. The basic kinetic theory of gases, as given in standard textbooks of physical chemistry, will be assumed.

1.2. Elastic collisions between inert gas molecules

1.2.1. Intermolecular potential energies

A conventional representation of the intermolecular potential energy curve is given by the Lennard-Jones 12-6 equation:

$$V(r) = 4\,\epsilon\,[(\sigma/r)^{12} - (\sigma/r)^{6}], \tag{1.1}$$

where $V(r)$ is the potential energy at separation r. This is shown graphically in Fig. 1.1. The second term, $(\sigma/r)^6$, represents the *long-range attractive forces*

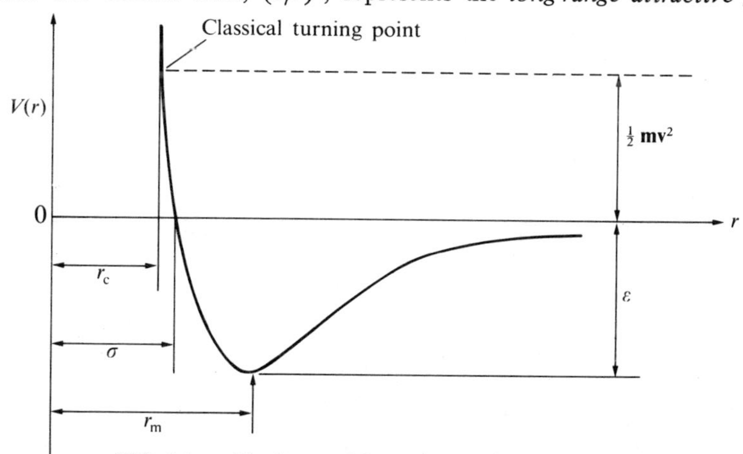

FIG. 1.1. The Lennard-Jones intermolecular potential

(London dispersion forces in the case of inert gases), and gives rise to the gentle decrease in $V(r)$ as the molecules first approach. The first term, $(\sigma/r)^{12}$, represents the *short-range repulsive forces* (due to overlap of electron orbitals), and produces the much steeper rise in $V(r)$ as they approach closer. ϵ is the depth of the 'potential well' at intermolecular distance r_m, where the repulsive forces take over from the attractive forces, and $V(r)$ is at its minimum. σ is the separation at zero energy, when $V(r) = 0$, sometimes loosely called the 'collision diameter'. Potentials other than the Lennard-Jones often give a better fit for particular systems, but all show the same general characteristics and use values for ϵ, r_m, and σ defined in the same way. A detailed treatment of the potentials available for both spherical and non-spherical (non-polar) molecules is given by Hirschfelder, Curtiss and Bird.[1] Inert gases and simple polyatomic gases give a fair approximation to the Lennard-Jones potential: a selection of parameters is given in Table 1.1.

TABLE 1.1

Collision parameters for non-polar molecules
Lennard-Jones 12-6 potential

Gas	Molecular weight	$\bar{v}$ at 273 K (m s^{-1})	σ (Å)	ϵ (cal mol^{-1})	Z (eqn 1.2) at 273 K (ns^{-1} atm^{-1})
He	4·0	1204·0	2·576	20·1	9·54
Ne	20·2	585·1	2·789	70·9	5·44
Ar	39·9	380·8	3·418	247	5·32
H_2	2·0	1692·0	2·915	75·5	17·2
N_2	28·0	454·2	3·749	159	8·68
O_2	32·0	425·1	3·541	175	6·35
CO	28·0	454·5	3·706	175	7·45
CH_4	16·0	600·6	3·796	286	10·8
CO_2	44·0	362·5	3·996	378	6·91
Cl_2	71·0	285·6	4·400	511	6·60
C_6H_6	78·1	272·2	5·270	874	9·02

Values of σ and ϵ from viscosity (ref. 1, Table 1-A).
(1 cal mol^{-1} = 4·184 J mol^{-1})

1.2.2. Collisions—the impact parameter

A head-on collision between two molecules may be crudely represented by a particle of mass m approaching from $r = \infty$ with speed v, and sliding without friction along the potential surface represented by the curve in Fig. 1.1. The total energy will remain constant at $\frac{1}{2}mv^2$ throughout the collision, and is represented by the horizontal dotted line. The particle will first accelerate as it runs down the attractive potential to r_m; it will then be retarded by the steep repulsive potential until it stops at a height $V(r) = \frac{1}{2}mv^2$. This is known as the *classical turning point*, and corresponds to a separation r_c smaller than σ. Since a collision between two inert gas molecules can only be an *elastic collision*, one in which the initial and final translational kinetic energies are the same, the motion will now be exactly reversed. It will be seen that the separation at the classical turning point will decrease appreciably as v increases, and is therefore temperature dependent, a fact which is important for transport properties.

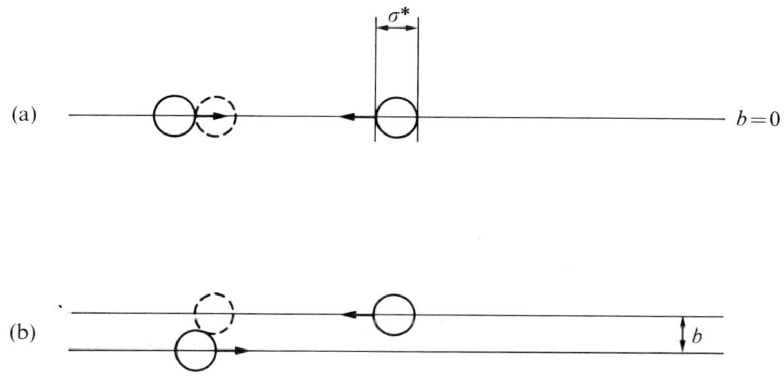

FIG. 1.2. The impact parameter

Most collisions are not head-on and are defined by an *impact parameter b*, which represents the distance of closest approach of the two molecular centres in the absence of any intermolecular potential. This is illustrated in Fig. 1.2(b). For a head-on collision (Fig. 1.2(a)) $b = 0$. On the hard-sphere model the maximum value b can have for a collision to occur is $b = \sigma^*$. The state of affairs for real molecules can be illustrated on a two-dimensional map of the potential surface surrounding the molecule. For a spherical molecule the one-dimensional model represented in Fig. 1.1 becomes a pair of 'mirror-image' Lennard-Jones curves, corresponding to the two sides of the molecule (Fig. 1.3(a)). The three-dimensional picture may be visualized as a steep conical tower, surrounded by a shallow moat and set in the middle of a flat plain. The two-dimensional map of this is given in Fig. 1.3(b). Inside the shaded circle of radius $r = r_m$ the potential is entirely repulsive. The outer circle represents the value of r where the attractive forces begin to operate. In the area between the two circles the potential is entirely attractive. Fig. 1.3(b) shows the trajectories of molecules at selected values of b. This diagram is only schematic, as the velocity of approach is important as well as the potential surface, but it makes clear the important point that for grazing collisions, when $b > r_m$, *only the attractive forces* operate. As b becomes smaller, both forces come into operation until the maximum perturbation by the intermolecular potential is achieved for $b = 0$, corresponding to the head-on collision discussed above.

In any kind of elastic collision it may be assumed that there is completely free interchange of translational energy between the collision partners, subject to the laws of conservation of energy and momentum. The rate of transfer of translational energy between molecules is thus effectively the gas kinetic collision rate, which we must now derive.

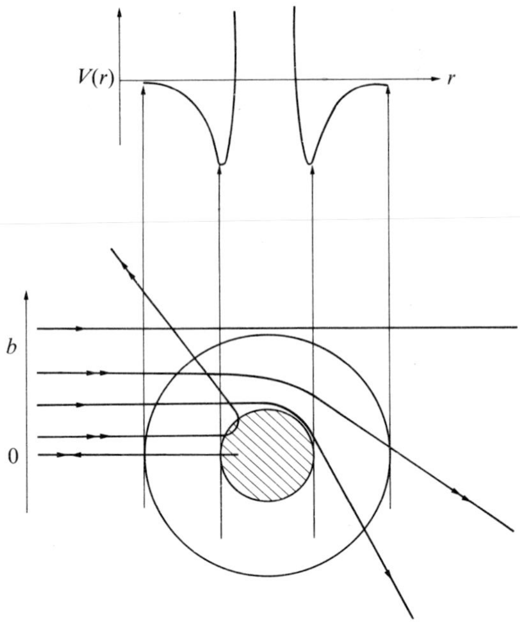

FIG. 1.3. Trajectories in molecular collisions

1.2.3. The gas kinetic collision rate

Elementary kinetic theory, in which molecules are treated as hard spheres with collision diameter σ^*, gives the number of collisions suffered by one molecule per second as

$$Z^* = 4n(\sigma^*)^2(\pi kT/m)^{\frac{1}{2}}$$

where n is the number of molecules in unit volume, and m the molecular mass. For real molecules following a Lennard-Jones type intermolecular potential, it is difficult to define exactly what is meant by a 'collision'. It is convenient in considering both energy transfer and chemical kinetics to define a 'gas kinetic collision rate', Z, as the number of occasions per second on which the molecular centres approach to within a certain distance.[2] Since both energy transfer and chemical reaction involve 'strong' collisions, in which the repulsive part of the potential plays a major role, this distance is usually taken as equal to σ (defined in § 1.2.1). The gas kinetic collision rate, the number of collisions made by one molecule per second, becomes

$$Z = 4n\,\sigma^2(\pi kT/m)^{\frac{1}{2}} \qquad (1.2)$$

Values for σ are obtained by fitting experimental viscosity or second virial coefficient measurements to the best available intermolecular potential.[1] For mixtures of two gases A and B, the number of collisions per second made by one molecular of species A with molecules of species B becomes

$$Z_{AB} = 2n_B \left(\frac{\sigma_A + \sigma_B}{2}\right)^2 \left[\frac{2\pi(m_A + m_B)kT}{m_A m_B}\right]^{\frac{1}{2}} = 2\sqrt{2}n_B\,\sigma_{AB}^2\,(\pi kT/\mu)^{\frac{1}{2}}$$

The quantity $\mu = m_A m_B/(m_A + m_B)$ is described as the *reduced mass* of the collision. This definition of Z and Z_{AB} will be used throughout this book, and is sometimes loosely called a 'hard sphere collision rate'. Values of Z calculated by eqn (1.2) are listed in Table 1.1.

An alternative method, used by some authors,[5,20] is to calculate collision rates directly from measured viscosities, using the formula: $Z = 5p/4\eta$ (where p is the pressure and η the coefficient of viscosity). Such collision rates are heavily weighted at lower temperatures in favour of grazing collisions involving only the attractive intermolecular forces.[2] These give rise to deflections in the path of the molecule, which are significant for the transport properties, but are too 'weak' to influence energy transfer. They may differ from rates calculated using eqn (1.2) by a factor of as much as two or three.

1.3. Inelastic collisions between diatomic molecules

1.3.1. Vibrational energy transfer

The simplest model to consider is a collinear collision between a vibrating diatomic molecule, AB, and a non-vibrating molecule, CD. This is likely to be the most efficient configuration for energy transfer, and is represented in Fig. 1.4. AB is assumed to be a simple harmonic oscillator, whose equilibrium

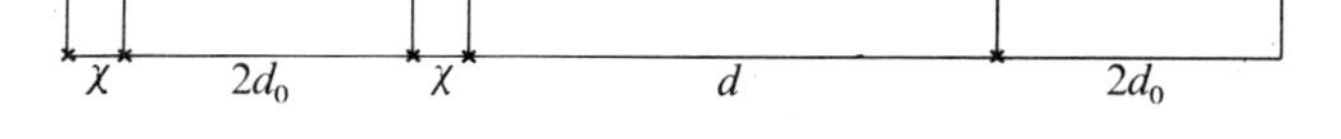

FIG. 1.4. Colinear collision between a vibrating and a non-vibrating diatomic molecule

internuclear distance is $2d_0$; CD a non-vibrating molecule with the same internuclear distance. r is the distance between the centres of mass of the two molecules. It will be a fair approximation, for this particular model, to assume that the variation of intermolecular potential energy, $V(r)$, with change in r during a collision depends only on the potential between the neighbouring atoms, B of molecule AB and C of molecule CD. The separation of B and C at any particular time will be given by $d = r - 2d_0 - \chi$, where χ is the displacement of the oscillator AB from equilibrium. $V(r)$ will vary with d during the collision in the same way as discussed in § 1.2.2, but in this case *both* r and χ

are varying with time. We thus have a changing external force or time-dependent perturbation, represented by the change in r, acting on a quantized periodic motion, represented by the variation of χ. The problem which interests us is whether this is likely to result in a transfer of energy between translation and vibration.

A simple physical interpretation is given by Ehrenfest's Adiabatic Principle:[3,5] if a changing force acts on a quantized periodic motion, the process will be *adiabatic* (no energy transfer) if the change of force is small during a period of the motion, and non-adiabatic (efficient energy transfer) if the change is large during this time. This may be expressed by defining an *adiabaticity parameter*,

$$\xi = \frac{\tau_c}{t_v}$$

where τ_c is the duration of the collision, and t_v the period of the oscillation. Large values of ξ correspond to adiabatic collisions, and the criterion for a non-adiabatic collision with efficient energy transfer is

$$\xi < 1 \quad \text{or} \quad \tau_c < t_v.$$

The duration of a collision may be regarded as $\tau_c = a/\bar{v}$, where a is the 'range' of the intermolecular forces and $\bar{v}$ the velocity of approach. The period of oscillation for a vibrating diatomic molecule is ν^{-1}, so that the criterion for efficient transfer becomes

$$a/\bar{v} < \nu^{-1} \quad \text{or} \quad \bar{v} > a\nu. \tag{1.3}$$

Taking the usual expression for the average velocity of approach at temperature T, this becomes

$$\left(\frac{8kT}{\pi m}\right)^{\frac{1}{2}} > a\nu.$$

Qualitatively, energy transfer will be favoured for collision partners with small mass, low vibration frequency, and steep intermolecular potential (corresponding to a small value of the range a). It will also become more efficient as the temperature rises. If energy transfer between vibration and translation does occur, the translational energy after the collision will differ from the translational energy before collision, and the collision is described as *inelastic* (in contrast to the elastic collisions described in § 1.2.2).

It is interesting to make a rough estimate of inequality (1.3) for an actual diatomic molecule whose vibrational mode is active at 300 K. Cl_2, whose collision parameters are given in Table 1.1, is a suitable example. The range of the intermolecular forces is of the order 2×10^{-8} cm and the average velocity is $2{\cdot}8 \times 10^4$ cm s^{-1}, so that the duration of the collision is $\tau_c = a/\bar{v} = 7{\cdot}15 \times 10^{-13}$ s. The fundamental vibrational frequency is 556 cm^{-1} = $1{\cdot}7 \times 10^{13}$ Hz, and the period of oscillation will be $t_v = \nu^{-1} = 5{\cdot}88 \times 10^{-14}$ s. τ_c is thus an order of magnitude larger than t_v, and the model predicts a predominantly adiabatic collision at room temperature, with only a very low probability of vibration-translation (V-T) occurring. Experiment shows that the average transfer probability per collision is $2{\cdot}2 \times 10^{-5}$ (see Fig. 3.5).

1.3.2. Rotational energy transfer

Fig. 1.5 represents the conditions for a collision between two rigid rotating molecules, AB and CD. It is at once clear that this is a far more complicated model to handle than the non-rotating collision shown in Fig. 1.4, as the varying interactions between all four atoms will need to be considered. But, again, there is essentially an external force, which changes during the collision, acting on a periodic motion, and Ehrenfest's Adiabatic Principle will apply.

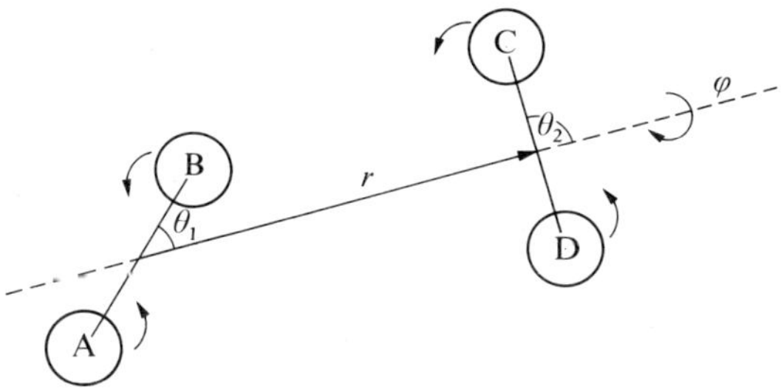

FIG. 1.5. Collision between two rotating diatomic molecules

If we consider the Cl_2 molecule, its average period of rotation at 300 K is 2×10^{-12} s, which is an order of magnitude *longer* than the collision duration, $7{\cdot}15 \times 10^{-13}$ s. This fulfils the criterion for efficient energy transfer, and it is found that rotation-translation (R-T) transfer occurs at almost every collision for most polyatomic molecules, except hydrogen and hydrides where rotation is much faster.

A further striking difference between V-T and R-T transfer is predicted by this simple model. While the vibration frequency of a molecule remains approximately constant with increasing vibrational energy (dropping slightly because of anharmonicity), the rotation frequency increases rapidly with increasing rotational energy. The resulting decrease in rotational period with rising temperature often outweighs the effect of increasing velocity of approach, so that the ratio (1.3) becomes less favourable and the efficiency of R-T transfer *decreases* with increasing temperature. (This is reflected in the quantum-mechanical treatment, which will be considered later, by the fact that, while the energy gap between adjacent vibrational levels, v and $(v - 1)$, remains approximately constant with increasing v, the gap between adjacent rotational levels, J and $(J - 1)$, increases rapidly with increasing J.)

1.4. Vibrational relaxation

1.4.1. The Landau-Teller theory

Landau and Teller[4,5] developed an elementary theory of vibrational energy transfer based on the model described in § 1.3.1. They assumed that only the repulsive part of the intermolecular potential (Fig. 1.1.) is steep enough to influence energy transfer, so that the long-range attractive potential can be neglected. If the repulsive potential is represented by a simple exponential $V(r) = V_0 \exp(-\alpha r)$ (replacing the Lennard-Jones term $4\epsilon(\sigma/r)^{12}$), the probability P_{10} for a $1 \to 0$ vibrational transfer occurring in a collision of velocity **v**, can be calculated by classical time-dependent perturbation theory on the assumption that P is proportional to the square of the Fourier component of the varying force at the oscillator frequency, ν. This assumes that all collisions have translational energy, $\frac{1}{2}m\mathrm{v}^2 \gg h\nu$, and gives P_{10} proportional to $\exp(-4\pi^2\nu/\alpha\mathrm{v})$. The distribution of molecular speeds is given by the Boltzmann equations, and the proportion of molecules with speed $\geqslant v$ is $\exp(-mv^2/2kT)$. Integration over the range of possible speeds gives the average probability of deactivation per collision at temperature T as:

$$P_{10} = \exp(-3(2\pi^4\mu\nu^2/\alpha^2 kT)^{\frac{1}{3}}). \tag{1.4}$$

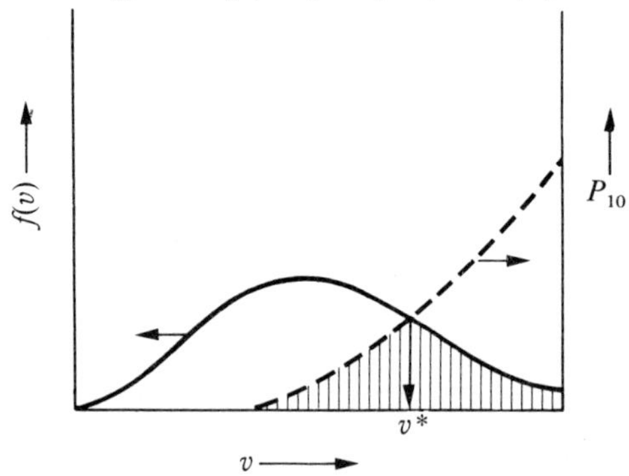

FIG. 1.6. Schematic representation of the overlap between the distribution of molecular speeds $f(v)$ (—), and the de-excitation probability P_{10} (- - -), as a function of molecular speed v.

The integration is represented graphically in Fig. 1.6, and it will be seen that, at each temperature, there is a speed v^* giving the highest transition probability. Eqn (1.4) is surprisingly good approximation to the results of more sophisticated quantum-mechanical treatments which will be described later. In particular, a great many non-polar molecules show a linear *Landau-Teller plot* of $\log P_{10}$ against $T^{\frac{1}{3}}$ (or $\log Z_{10}$ against $T^{-\frac{1}{3}}$) over a large temperature range.

1.4.2. The vibrational relaxation time

Theoretical models, such as the Landau-Teller, give values of P_{10}, the average probability per collision for a v $(1 \to 0)$ transition to occur. In order to make comparison with experimental measurements this must be related to a transfer *rate*. The average number of collisions required for a molecule in the $v=1$ state to deactivate will be $Z_{10} = P_{10}^{-1}$, which is defined as the *collision number* for a v $(1 \to 0)$ transfer. The gas kinetic collision rate (1.2) is $Z = 4n\,\sigma^2\,(\pi kT/m)^{\frac{1}{2}}$ collisions made by one molecule per second. The average time required for a molecule in state $v=1$ to deactivate is thus:

$$\beta = Z_{10}/Z = Z_{10}/4n\sigma^2(\pi kT/m)^{\frac{1}{2}} \qquad (1.5)$$

β is known as the vibrational *relaxation time*, and is the characteristic time for establishment of equilibrium between translational motion and the intramolecular vibration. Since Z is proportional to the molecular concentration n, and hence to pressure, β will vary inversely with pressure. For convenience relaxation times are always referred to a pressure of 1 atm.

Eqn (1.5) is an approximation. A v (1-0) transfer can only occur when a molecule in the v=1 state collides with another molecule in the v=0 state. We have tacitly assumed that all the other n molecules are in this state. In fact there will be a Boltzmann distribution of molecules between the two levels. Further, the process is reversible and, by the principle of detailed balancing, P_{01} for the reverse transfer must be given by

$$P_{01} = P_{10}\exp(-h\nu/kT).$$

Taking these factors into account, the correct form of relation (1.5) for a simple harmonic oscillator is:[5]

$$Z_{10} = Z\beta\,\{1-\exp(-h\nu/kT)\} = P_{10}^{-1}. \qquad (1.6)$$

Experimental measurements usually give a relaxation time β from which P_{10} and Z_{10} may be calculated by eqn (1.6). The collision number Z_{10} is the most convenient parameter to use.

An alternative way of expressing measurements is in terms of a *rate constant* k for energy transfer, defined by the relation

$$kp = \beta^{-1} \quad \text{or} \quad k = \beta^{-1}p^{-1}, \qquad (1.7)$$

p being the pressure. (k is usually expressed in the units s^{-1} atm^{-1}, or s^{-1} $torr^{-1}$). The transfer rate may also be expressed in terms of a hypothetical *energy transfer collision cross-section* Q, defined by the relation

$$kp = Z(Q/\pi\sigma^2) = \beta^{-1}. \qquad (1.8)$$

($\pi\sigma^2$ is the gas kinetic collision cross-section). For a single gas, using eqn (1.2) for Z, we have

$$kp = 4n(kT/\pi m)^{\frac{1}{2}}Q.$$

Both k and Q may be related to P_{10} by eqns (1.6), (1.7), and (1.8). For experimental conditions where $h\nu \gg kT$ (e.g. most experiments involving diatomic molecules), eqn (1.5) replaces (1.6), so that $Q/\pi\sigma^2 = P_{10}$, and the cross-section is a direct measure of transfer probability.

1.5. Rotational relaxation

The model for a collision between rotating molecules discussed in § 1.3.2 showed that any quantitative treatment for R-T transfer involves complicated geometrical factors. A further complication arises out of the fact that for most molecules the separation between lower rotational quantum levels is so small that, except at low temperatures, a wide range of rotational energy levels will be populated under equilibrium conditions. This is in marked contrast with V-T transfer, where the majority of molecules are in the $v = 0$ level, and only v ($0 \rightarrow 1$) transitions are important. R-T transfer can be treated by a purely classical model, and Parker[21] showed that rotational relaxation can be characterized by a single rotational relaxation time, β_R, involving the whole of the rotational energy of the molecule, and corresponding to a rotational collision number Z_R defined as in § 1.4.2. But neither β_R nor Z_R can be related to a single rotational transition, $J_n \rightarrow J_m$. Parker found a complicated expression for homonuclear diatomic molecules, which predicts an increase in Z_R with rising temperature, and gives a value $Z_R \sim 5{\cdot}0$ for Cl_2 at 300 K, which is in fair agreement with experiment. More recent classical and quantum-mechanical treatments will be considered in Chapter 5.

1.6. Collisions between polar molecules–dimerization

Collisions between polar molecules differ from the model discussed in §§ 1.2 and 1.3 since the intermolecular potential $V(r)$ becomes strongly dependent on the relative orientation of the approaching molecular dipoles. The potential is now the sum of four different contributions: (i) the short-range repulsion, (ii) the London dispersion forces, which depend only weakly on orientation, (iii) the angle dependent electrostatic contribution due to interaction of the dipoles, and (iv) a dipole-induced dipole term. Calculations of the bulk properties of the gas are usually based on the Stockmayer 12-6-3 potential:[1]

$$V(r, \theta_1, \theta_2, \phi_2 - \phi_1) = 4\epsilon[(\sigma/r)^{12} - (\sigma/r)^6] - \frac{\mu^2}{r^3} g(\theta_1, \theta_2, \phi_2 - \phi_1). \quad (1.9)$$

This is essentially a combination of a Lennard-Jones type potential, in which σ and ϵ represent respectively the distance of separation at zero energy and the maximum energy of interaction in the *absence* of dipole forces, with a third term, involving μ^2/r^3, representing the angle dependent dipole interaction. μ is the dipole moment, θ_1 and θ_2 are the angles made by the dipoles with the line joining their centres, and ϕ the angle between the planes which pass through the line of centres and contain the two axes (as shown in Fig. 1.5). The maximum energy of dipole interaction, $\Delta U = -2\mu^2/r_0^3$, will be effective over a restricted range of configurations in which the dipoles are in the end-to-end position. The value of r_0 will be less than that of r_m (Fig. 1.1), and will depend on the strength of the dipole-dipole interaction. Values of these parameters for some polar molecules, calculated by Rowlinson[6] to fit experimental second virial coefficients, are given on Table 1.2.

Table 1.2 shows that the values of ΔU for the more strongly polar molecules are several times larger than the values of ϵ. In terms of the pictorial account of the potential energy surface in § 1.2.3, the shallow moat surrounding the conical tower now has a very steep dip on one side. When two molecules undergo a collision in this orientation, the intermolecular potential will differ

TABLE 1.2.

Potential energy parameters for polar gases.
Stockmayer 12-6-3 potential

Gas	$\mu \times 10^{18}$ (esu cm)	σ (Å)	ϵ (cal mol^{-1})	$-\Delta U$ (cal mol^{-1})	$-\Delta U_{exp}$ (cal mol^{-1})
$CHCl_3$	1·05	2·99	2110	860	–
CH_3Cl	1·89	3·44	750	2020	2390
CH_3COCH_3	2·74	3·76	1030	3280	–
CH_3OH	1·66	2·40	1260	4660	3300 (a)
NH_3	1·47	2·58	630	2960	3680
H_2O	1·83	2·65	760	4440	4980
CH_3CN	3·50	4·02	800	4640	4500
CH_3CHO	2·70	3·68	530	3680	3790
SO_2	1·63	4·04 (b)	690 (b)	–	4750 (c)

Data from ref. 6, except:
(a) Kell, G. S. and McLaurin, G. E. (1969). *J. chem. Phys.* **51**, 4345.
(b) Monchick, L. and Mason, E. A. (1961). *J. chem. Phys.* **35**, 1677.
(c) Pemberton, D. (1959). Chemistry Part II Thesis, Oxford.

(1 cal mol^{-1} = 4·184 J mol^{-1})

from the Lennard-Jones potential in the way shown in Fig. 1.7. The gradients of both attractive and repulsive parts become much steeper, in addition to the increased depth of potential well, and this leads to greatly increased efficiency of energy transfer. This favourable orientation will be *preferred* for collisions at lower temperatures where $\Delta U > RT$, and in this region the vibrational transition probabilities will show a decrease as the temperature rises.[68,69] At higher temperatures, where $\Delta U < RT$, the effect ceases to be important and the positive temperature coefficient predicted by Landau and Teller (§ 1.4.1) will take over. The plot of P_{10} against T for polar molecules usually shows a pronounced minimum in the region of 600 K (see § 3.3.2).

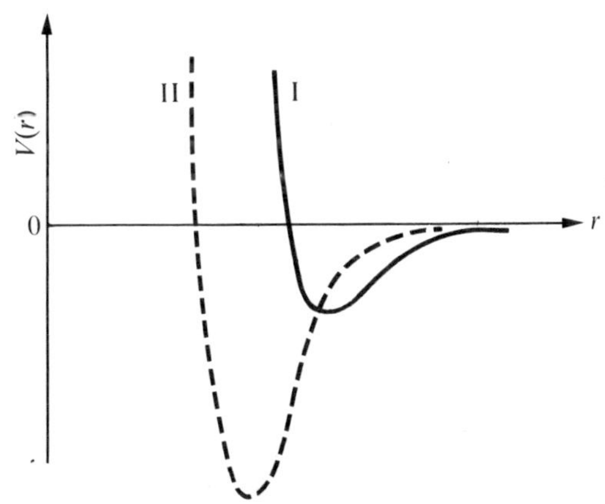

FIG. 1.7. Intermolecular potentials: I, Lennard-Jones, II, oriented dipole-dipole

For molecules where $\Delta U \geqslant 3000$ cal mol^{-1}, this is some 5 times larger than RT at 300 K so that there will be a significant chance of a stable dimer being formed when two molecules collide in the appropriate orientation. An alternative to the Stockmayer treatment of second virial coefficients for strongly polar molecules treats their deviation from the principle of corresponding states (which is closely followed by non-polar molecules) as entirely due to a reversible dimerization process which follows the law of mass action,[6,7,8] and leads to simple calculation of a heat of dimerization, ΔU_{exp}. Some values of ΔU_{exp} are shown in Table 1.2 and show fair agreement with the values of ΔU calculated by Rowlinson.[6] For molecules such as NH_3 and H_2O, where hydrogen-bonding is possible, the values of r_0 and ΔU are in fair agreement with the accepted values for hydrogen bonds.[9] Independent experimental evidence for the presence of dimers in the gas phase is given by spectroscopic measurements on CH_3OH,[10] and on NO,[11,13] and by measurements of transport properties of H_2O and NH_3.[12,19] Dimerization due to quadrupole-quadrupole forces has been inferred from second virial coefficient measurements on CO_2,[14,15] giving a value $\Delta U_{exp} = -2980$ cal mol^{-1}. These large interaction energies are comparable with the quantum of vibrational energy transferred in an inelastic collision. The energy involved in a $v = 0 \rightarrow 1$ transition, where $\nu = 1000$ cm^{-1}, is only 2860 cal mol^{-1}, so that the energy of dimerization is, in principle, sufficient to produce a $v = 0 \rightarrow 1$ transition even in a collision with zero velocity of approach. Dimerizing collisions would be expected to give highly efficient energy transfer, which will again show a strong negative temperature-dependence.

It is interesting that evidence for appreciable dimerization in *non-polar* gases at low temperatures, where $\epsilon > RT$, has been inferred from measurements of second virial coefficients and of transport properties.[16] Their presence has been confirmed by mass spectrometry[17] and by optical spectroscopic measurements.[18] Energy transfer measurements have recently been extended to much lower temperatures for such gases, and show that, under these conditions, the accelerating effect of the attractive potential plays a significant role in promoting energy transfer. This produces some departure from the Landau-Teller temperature-dependence, which is based on a purely repulsive interaction.

2

EXPERIMENTAL INVESTIGATION OF ENERGY TRANSFER

2.1. Introduction

When a gas is at equilibrium, the distribution of energy between translation, vibration, and rotation is governed by the Boltzmann relations. Another way of expressing this is to say that the translational, vibrational, and rotational *temperatures* are equal: $T_{TR} = T_{VIB} = T_{ROT}$. The rate of transfer of energy between modes may be followed by inducing a small change in one of these temperatures and observing the rate of return to equilibrium. Provided the change is small, this will be a relaxation process, governed by an equation (e.g. for vibrational relaxation)

$$\frac{dT_{VIB}}{dT} = \frac{1}{\beta}(T_{VIB} - T^*_{VIB}), \qquad (2.1)$$

where T^*_{VIB} is the final equilibrium vibrational temperature. β is the vibrational relaxation time, and corresponds exactly with the model discussed in § 1.4.2. The methods for producing a disturbance of the Boltzmann equilibrium fall into two main classes:

(a) When a gas undergoes adiabatic compression, which may be visualized as due to the mechanical action of a frictionless piston, the translational energy alone is directly influenced. In a sudden compression the whole energy goes initially to increase T_{TR}, and this followed by relaxation due to T-V and T-R energy transfer until equilibrium between the three modes is re-established. Such rapid compressions are produced by ultrasonic sound or by the passage of shock-fronts. These form the basis of the classical methods for determining relaxation times, and will be discussed in §§ 2.2 and 2.3.

(b) When a gas absorbs electromagnetic radiation, selected vibrational and rotational modes may be excited. This results in an increase of T_{VIB} and/or T_{ROT}, but has no direct effect on T_{TR}. Relaxation to equilibrium by V-T or R-T transfer may be followed either by monitoring the decrease in T_{VIB} or T_{ROT} by spectroscopic methods, or by monitoring the accompanying increase in T_{TR}. These form the basis of the newer methods for measuring relaxation times, and are discussed in §§ 2.4, 2.5, and 2.6. Gases with non-equilibrium values of T_{VIB} and T_{ROT} are also sometimes produced in fast chemical reactions, and the relaxation may be followed by similar methods (§ 2.7).

There are, in addition, two methods for investigating energy transfer which do not involve temperature relaxation. The first is the molecular beam technique (§ 2.8), which yields direct information about what happens in a collision between two individual molecules. The second involves measurement of the ratio between thermal conductivity and viscosity of a gas. A development of kinetic theory due to Mason and Monchick[22] relates this to the collision number for rotational energy transfer. This method yields inaccurate values of Z_R, but the phenomenon has interesting theoretical implications, and will be discussed in § 2.9.

2.2. Acoustic methods

2.2.1. Theory of dispersion and absorption

The passage of a sound wave through a gas involves rapidly alternating adiabatic compression and rarefaction. The adiabatic compressibility of a gas is a function of γ, the ratio of the specific heats. The classical expression for the speed, W, of sound in a perfect gas of molecular weight, M, is:

$$W^2 = \gamma(RT/M) = (RT/M)(1 + R/C_V). \tag{2.2}$$

For a monatomic gas, where the heat capacity involves only translational energy, W is independent of sound oscillation frequency (except at very high frequencies or very low pressures where a classical viscothermal dispersion occurs). For a relaxing polyatomic gas this is no longer so. At sound frequencies where the period of the compression-rarefaction cycle becomes comparable with the relaxation time, the internal temperature of the gas lags behind the translational temperature throughout the cycle, and the effective values of C_V and β become frequency dependent. This results in a *dispersion zone* of frequencies over which W changes from a value W_0, corresponding to the static heat content, C_{V_0}, at lower frequencies to a higher value W_∞, corresponding to $C_{V_\infty} = (C_{V_0} - C_1)$ at higher frequencies where the relaxing mode ceases entirely to make its usual contribution, C_1, to the effective heat capacity. Speed *dispersion* is accompanied by a non-classical *absorption* of sound, because the internal temperature lag causes dissipation of energy.

The full mathematical theory of dispersion and absorption is given in standard works.[5,20,23] This involves alternative definitions of the relaxation times for different conditions, which will be explained here, as they sometimes appear in experimental papers. The definition of β given in §§ 1.4.2 and 2.1 corresponds to an *isothermal* relaxation process in which T_{TR} is maintained constant by contact with a heat bath. Such a process is represented in Fig. 2.1(a). The translational temperature T_{TR}, represented by the full line, is suddenly increased by an amount ΔT_{TR}, and is then maintained constant at the new temperature by a heat bath. The internal temperature T_I relaxes along the dashed line, with relaxation time β. T_{TR}, pressure, and volume all remain constant during the relaxation, and we may write for the isothermal relaxation time: $\beta = \tau_{VT} = \tau_{pT}$. An *adiabatic* relaxation process under conditions of constant *pressure* is represented in Fig. 2.1(b). This is *isentropic*, and T_{TR} falls as T_I rises until the two curves meet; pressure remains constant and volume decreases. This process corresponds to an adiabatic constant pressure relaxation time,

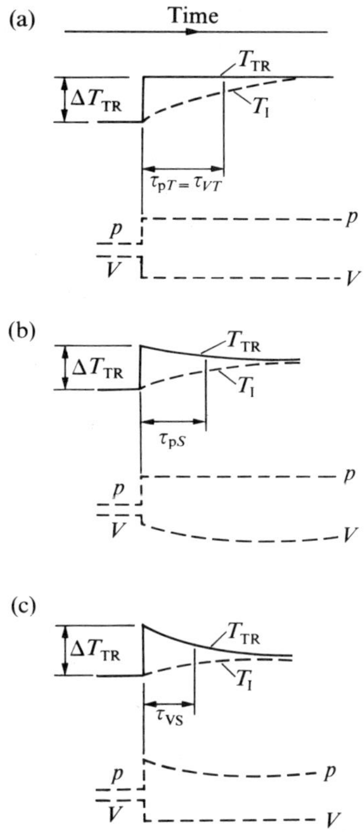

FIG. 2.1. Temperature relaxation processes. (a) Isothermal; (b) adiabatic, constant pressure; (c) adiabatic, constant volume

τ_{pS}. Adiabatic relaxation under conditions of constant *volume* is shown in Fig. 2.1(c), and corresponds to an adiabatic constant volume relaxation time τ_{VS}. It may be seen that

$$(\beta = \tau_{VT} = \tau_{pT}) > \tau_{pS} > \tau_{VS},$$

and the three relaxation times are related by the equations

$$\tau_{pS} = \tau_{pT}(C_{p_\infty} - C_{p_0})$$
$$\tau_{VS} = \tau_{pT}(C_{V_\infty} - C_{V_0}). \qquad (2.3)$$

Provided these relations are borne in mind, any relaxation process may be expressed in terms of any of the three relaxation times, with appropriate modification of the equations. It is convenient always to use the isothermal relaxation time β, which is directly related to the molecular energy transfer process (and is sometimes called the 'true relaxation time),[5] although this gives less elegant expressions for acoustic phenomena, which involve adiabatic processes.

The variation of sound speed with angular frequency ω (=$2\pi f$) is given by

$$W^2 = RT/M\left[1+\frac{R(C_{V_0}+C_{V_\infty}\omega^2\beta^2)}{(C_{V_0}^2+C_{V_\infty}^2\omega^2\beta^2)}\right]. \tag{2.4}$$

This is usually represented graphically by plotting $(W/W_0)^2$ against $\log \beta\omega$, as shown in Fig. 2.2(a). It will be noted that the point of inflection of the curve corresponds to $\omega\tau_{VS} = 1$. An alternative presentation is to calculate an 'effective' value of C_V, $C_{V_{\text{eff}}}$, from the measured value of W by the equation

$$W^2 = RT/M(1+R/C_{V_{\text{eff}}}). \tag{2.5}$$

Combination of eqns 2.4 and 2.5 then predicts a plot of $C_{V_{\text{eff}}}$ against $\log \beta\omega$ as shown in Fig. 2.2(b). This method of presentation has the advantage that C_1, the

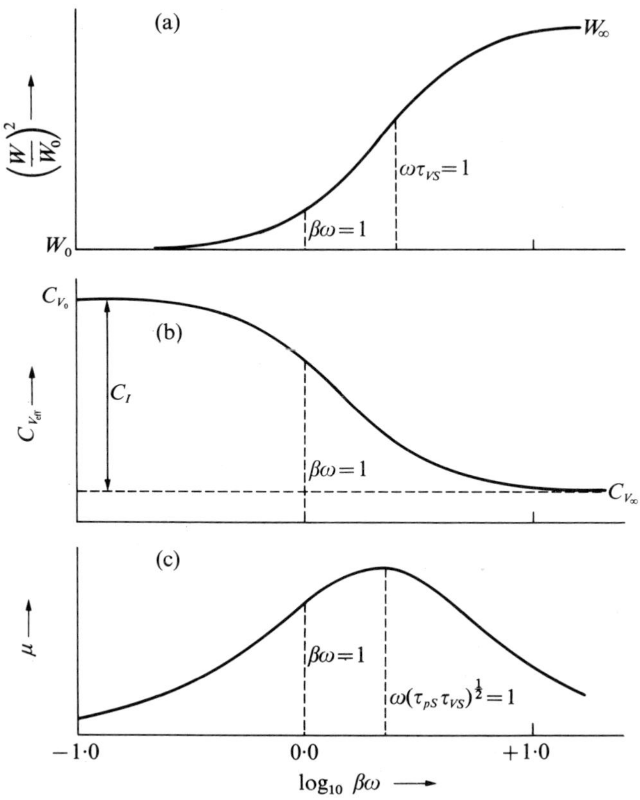

FIG. 2.2. Acoustic dispersion and absorption curves

contribution of the relaxing mode to the total specific heat, may be read directly off the plot. The variation of the *absorption coefficient* per wavelength, μ, with cyclic frequency is given by

$$\mu = \frac{2\pi\omega C_{V_\infty}\beta}{C_{V_0}}\left[\frac{W^2/W_0{}^2-1}{(W^2/W_0{}^2)(C_{V_\infty}\omega\beta/C_{V_0})^2+1}\right]. \tag{2.6}$$

The plot of μ against log $\beta\omega$ is shown in Fig. 2.2(c): maximum absorption occurs where $\omega\ (\tau_{pS}\ \tau_{VS})^{\frac{1}{2}} = 1$. Experimental values of are obtained by making measurements of either sound speed or sound absorption at varying frequencies, and fitting the results to the appropriate dispersion or absorption curve. Since β is inversely proportional to pressure, doubling the pressure is exactly equivalent to halving the frequency in the context of eqns. (2.4) and (2.6) and it is usually convenient to make a series of measurements at constant frequency and varying pressure, and to plot $\log(f/p)$ as the abscissa.

Sound speed measurements, need careful correction for the effect of gas imperfection. For a real gas W, C_p, and C_V all vary with pressure in a manner depending on the second virial coefficient B,[24] and must be corrected to give the ideal values corresponding to zero pressure.[23] The variation of W with p for a non-dispersing real gas is given by

$$W^2 = (\gamma/M)(RT+2Bp). \tag{2.7}$$

This is used to calculate a value of γ from the measured value of W. An ideal value of $C^*_{V\mathrm{eff}}$ is calculated from γ using the equation

$$C^*_{V\mathrm{eff}} = \frac{R}{\gamma-1} + p\left[\frac{2\gamma}{\gamma-1}\,\frac{\mathrm{d}B}{\mathrm{d}T} + T\frac{\mathrm{d}^2B}{\mathrm{d}T^2}\right], \tag{2.8}$$

and this plotted against log (f/p). Alternatively, the ideal value of W^* may be calculated from C^*_V by eqn (2.5). These corrections are vitally important, since for polyatomic gases their magnitude is comparable to that of the velocity dispersion itself. They require accurate knowledge of second virial coefficients over a range of temperatures: a useful compilation of these has been made by Dymond and Smith.[25]

Sound absorption measurements require correction for a classical viscothermal absorption, due to the influence of viscosity and thermal conductivity, in addition to the correction of W in eqn (2.6) for gas imperfection. The classical absorption per wavelength is given by

$$\mu_{\mathrm{class}} = (2\pi^2/\gamma_0)\left[\frac{4}{3}\eta + (\gamma_0-1)\frac{kM}{C_{p_0}}\right]\frac{f^2}{p}, \tag{2.9}$$

where η is the coefficient of viscosity and k the thermal conductivity.[23] This must be added to the absorption due to relaxation. For values of $f/p < 10^6$ Hz atm^{-1}, μ_{class} may be neglected; at higher values it rapidly becomes dominant, as shown in Fig. 2.3. These corrections are also vitally important in the

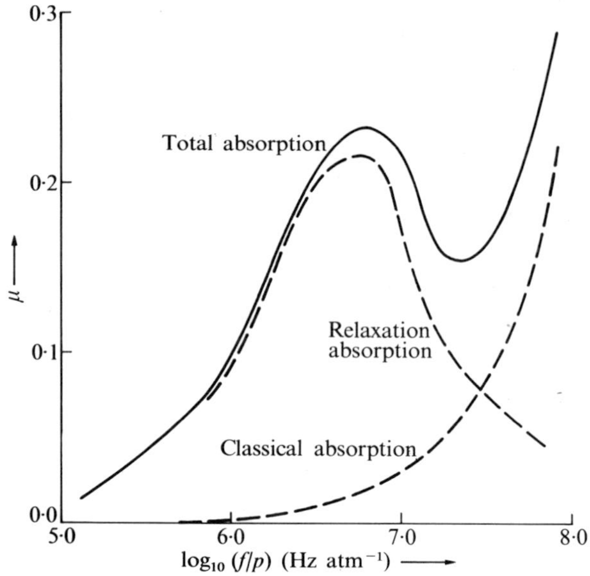

FIG. 2.3. Absorption curves for C_2D_2 (ref. 29), β (vibrational) = 61 ns

interpretation of absorption measurements, and require accurate values of thermal conductivity and viscosity coefficients. A full theoretical discussion of the behaviour of sound velocity and absorption at high values of f/p is given by Greenspan.[26]

The discussion above has been conducted entirely in terms of a *single* relaxation process, corresponding to a single value of β. In practice all gases show two separate relaxation processes for vibration and rotation, with different values of β, and some polyatomic gases also show different relaxation times for different vibrational modes. Rotational relaxation times are commonly several orders of magnitude shorter than vibrational, and the complete dispersion and absorption curves for a diatomic molecule take the form shown in Fig. 2.4. Here each relaxation process can be treated independently, using appropriate values of C_{V_0} and C_{V_∞}. For multiple vibrational relaxation, involving two separate vibrational specific heat contributions, C_1 and C_2, where $\beta_1/\beta_2 < 10$, more sophisticated treatment is necessary, based on the energy and temperature relaxation equations developed by Tanczos.[27,28] These involve adiabatic constant volume relaxation times τ_{VS_1} and τ_{VS_2}, associated with relaxation 'constants' D_1 and D_2, defined by the relation

$$D_1 = RC_1'/C_{p_0}(C_{V_0} - C_s). \tag{2.10}$$

C_s is the total (static) vibrational specific heat contribution, and C_1' is an 'apparent specific heat contribution' which is not exactly equal to C_1.

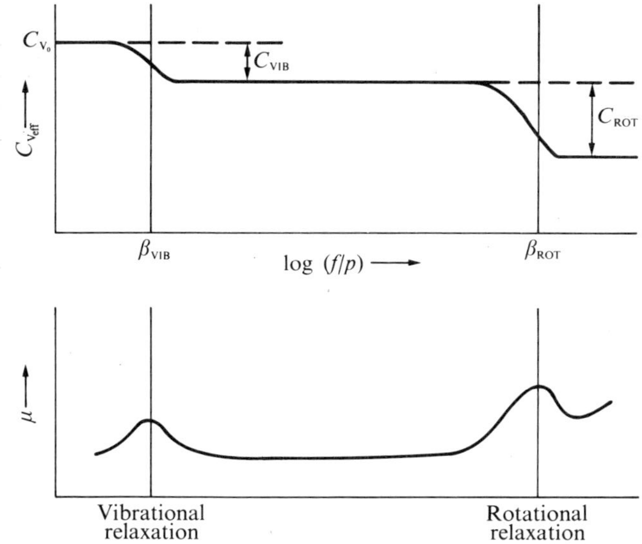

FIG. 2.4. Dispersion and absorption curves for a diatomic molecule (schematic)

2.2.2 Acoustic instruments

Measurements of sound speed at ultrasonic frequencies are usually made by an acoustic interferometer. An example of this apparatus is shown in Fig. 2.5. An optically flat, X-cut piezo-quartz crystal with gold-plated faces is set into oscillation by an appropriate electrical circuit, which is coupled to an accurate means of measuring electrical power consumption. A reflector, consisting of a bronze piston with an optically flat head set parallel to the oscillating face of the quartz, is moved slowly towards or away from the quartz by a micrometer screw. The electrical power consumption shows successive fluctuations as the distance between quartz and reflector varies between positions of resonance and non-resonance of the gas column. Measurement of the distance between resonance positions gives a value for $\lambda/2$, and if f is the frequency of the quartz, $W = f\lambda$ can be calculated. Frequencies ranging between 100 kHz and 10 MHz are used, giving wavelengths ranging between 0·5 and 0·005 cm. As the wavelength is always small compared to the interferometer dimensions, 'tube effects' are unimportant, and wavelengths can usually be measured correct to 1 in 1000. It is most convenient to drive the quartz at its resonant frequency (which must be accurately monochromatic), and to scan the dispersion zone by varying the gas pressure.

Measurements of sound absorption are made in instruments similar to the interferometer described above, but with a receiving transducer substituted for the reflector, so that the strengths of transmitted and received signals can be directly compared. It is best to use a pulse technique, and quartz crystals

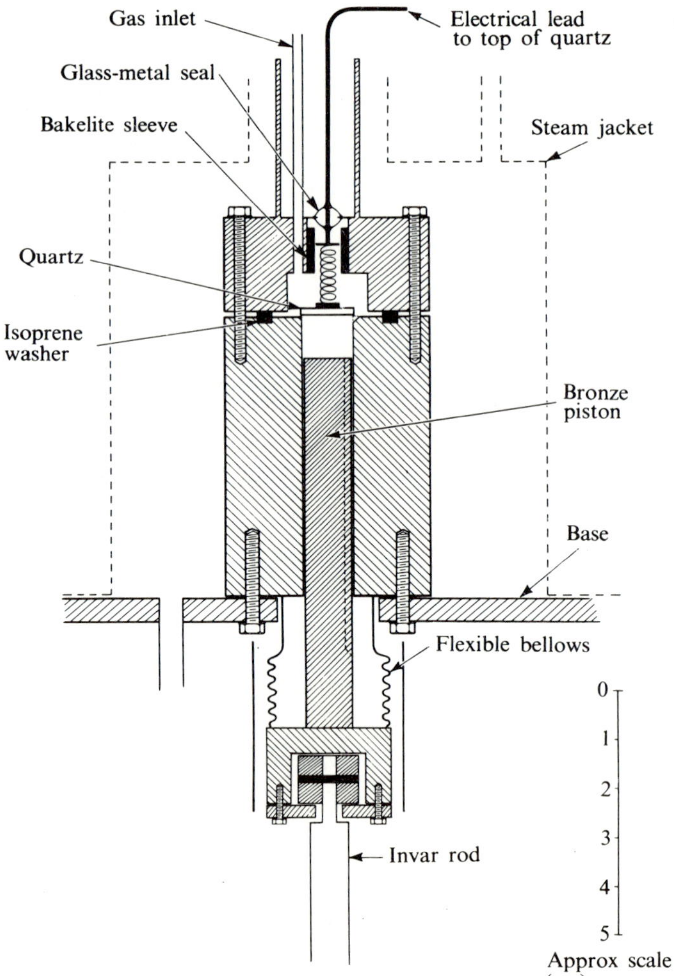

FIG. 2.5. 4 MHz acoustic interferometer[30]

are often replaced by condenser transducers, where the oscillating surface is a thin aluminized Melinex plastic sheet stretched over a flat metal backing-plate. An example of this apparatus is shown in Fig. 2.6. It is capable of operating over a continuous range of frequencies, so that there is no need to match accurately the two transducers before use. It can also operate at much lower gas pressures, and correspondingly higher f/p values, than quartz. This more sophisticated apparatus may also be used for speed measurements, the wavelength being determined by monitoring the phases of transmitted and received sound pulses on an oscilloscope, while the separation of the transducers is slowly varied. Full details are given by Stretton.[23]

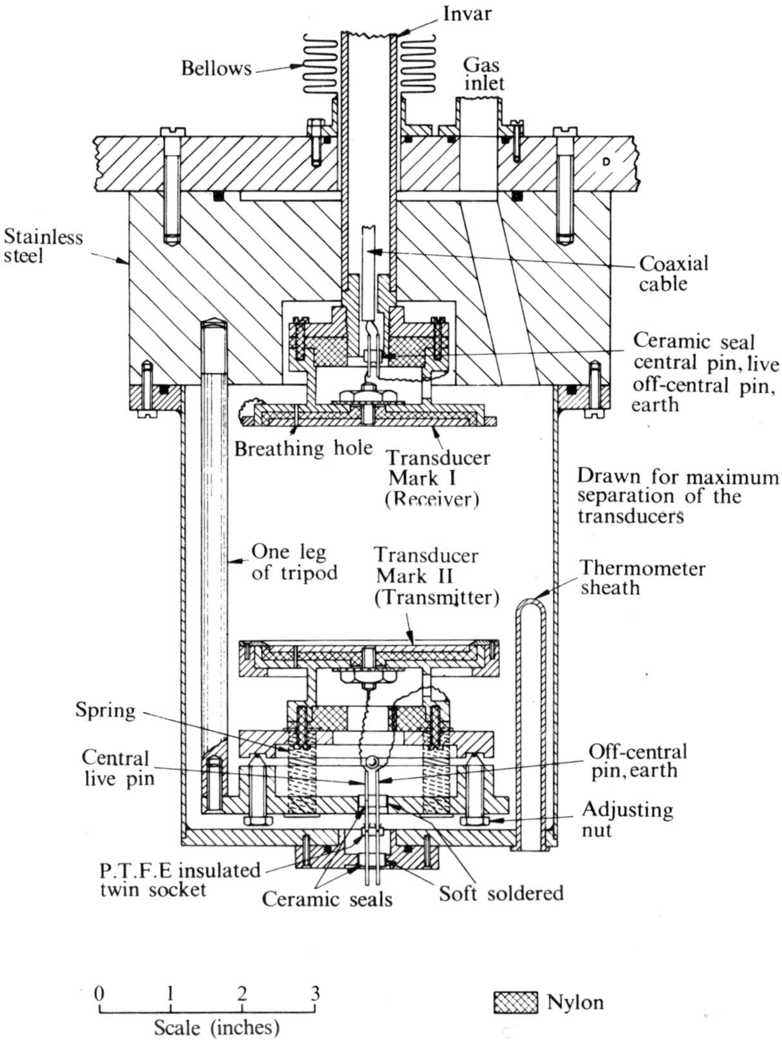

FIG. 2.6. Double condenser-transducer apparatus of Stretton[23]

An interesting variant of the acoustic interferometer is to determine the wavelength by optical diffraction. Travelling ultrasonic waves are set up in an acoustic chamber by a circular, 4 cm diameter, X-cut quartz crystal of frequency ~1 MHz. A parallel beam of monochromatic light from a slit crosses the sound field perpendicularly, and the resulting interference patterns are recorded and measured on a photographic plate.[31,32,33] The apparatus is calibrated by using a standard gas, and yields results which differ by less than 1 in 1000 from those obtained by the apparatus shown in Fig. 2.6, and from speeds calculated from accurate specific heat data.

A large part of the available data on relaxation times of polyatomic molecules ranging from 1 ns to 6 μs has been obtained by the type of experiment described above. It is probably the most accurate technique available, and has the advantage that it requires only a small sample of gas, whose purity (which is critically important, cf. § 3.5.4) can be easily controlled. It is not suitable for measuring relaxation times longer than ~6 μs, which would require sonic frequencies lower than 100 kHz, where the sound wavelength becomes comparable with apparatus dimensions. Its other disadvantage is that it cannot be used at high temperatures (where the vibrational modes of many diatomic molecules first become active); quartz transducers will not operate at temperatures above 350-400°C, the condenser-transducer cannot be used above 100°C, and no other suitable transducers are available. The method is therefore restricted to investigation of T-R and T-V transfers involving low-lying energy levels of polyatomic molecules.

An acoustic technique which is suitable for making absorption measurements at lower frequencies (down to 100 Hz) and at high temperatures is the reverberation method; an apparatus is shown in Fig. 2.7. The resonator cylinder

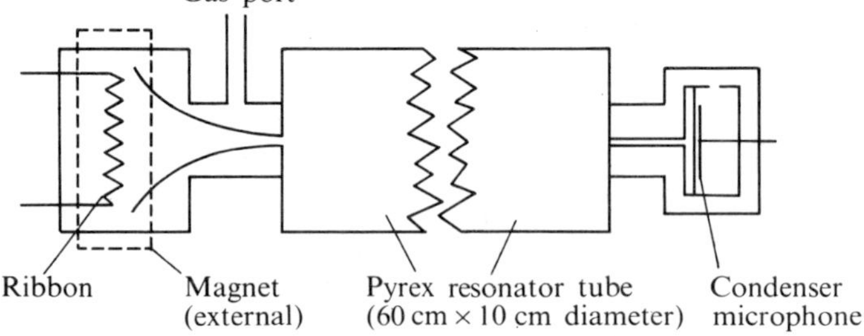

FIG. 2.7. Diagram of the resonator of Edmonds and Lamb[34]

containing the gas is excited intermittently, at one of its eigenfrequencies, by a loud-speaker. The exponential decay of the excess pressure in the resonator after the excitation has ceased is detected by a condenser-transducer coupled to the cylinder through a probe tube. The theory is fully discussed by Edmonds and Lamb.[34,35] Apparatus of this type has been used to measure a vibrational relaxation time β~16 ms for oxygen at 30°C.[36,37] Lukasik and Young[38] measured vibrational relaxation times β~0·8 ms for nitrogen at temperatures up to 1186°C by employing transducers external to the hot resonant cavity. A detailed discussion of these and other acoustic methods will be found in ref. (23).

2.3. Shock-tube methods

2.3.1. Formation and properties of shock fronts

The speed of sound in a gas increases with rise in temperature (eqn 2.2). This means that in a sound wave, regarded as a periodic compression-rarefaction,

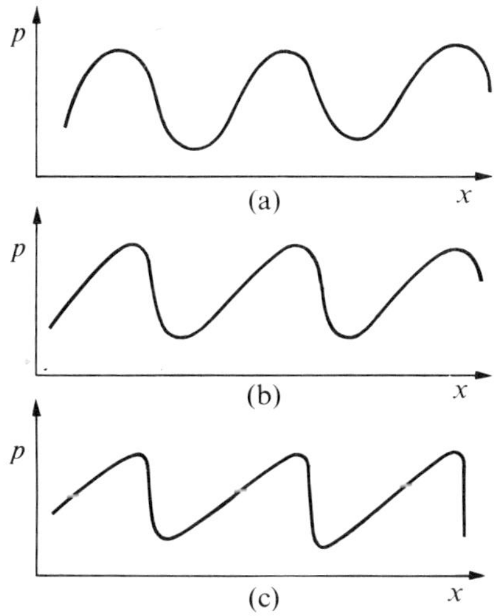

FIG. 2.8. The formation of a shock wave

the crest of the compression wave, which is at a higher temperature than the trough, will move with a higher speed. Fig. 2.8(a) represents a sinusoidal wave travelling in the positive x-direction. After some time this will assume the shape shown in Fig. 2.8(b). Eventually the crest will overtake the trough, and the wave will assume the saw-tooth shape of Fig. 2.8(c). There is now an abrupt discontinuity at the front of the saw-tooth: this is called a shock front. The process is analogous to the formation of breakers when an ocean swell reaches a beach. The speed of a shallow-water wave is proportional to the depth of the water, so that the top of the wave travels at a higher speed than the trough and the leading edge eventually becomes vertical. There is then a violent discontinuity and the wave 'breaks'. In normal sound propagation the effect is much too small for a shock front to occur before the wave motion dies out. But if a gas is subjected to sudden massive acceleration by a moving piston, a shock front is formed after a very short distance. This discontinuity is a sudden adiabatic compression which is now irreversible and is no longer isentropic.

Shock waves are produced in the laboratory by the sudden bursting of a diaphragm separating two gases at different pressures: the interface acts as the moving piston described above. A schematic shock tube is shown in Fig. 2.9(a). It would commonly be a metal tube of diameter 5-10 cm and total length 5-6 m. A shaped metal diaphragm separates the high-pressure gas (p_4) from the low-pressure gas (p_1). The diaphragm is burst mechanically, and a compression wave is formed in the low-pressure gas, which rapidly steepens to form a shock front. The shock front moves into the low-pressure gas at a speed W_s, which is greater than the speed of sound in the gas, W_1. The ratio $W_s/W_1 = M_I$ is known as the Mach number: it depends on the pressure ratio p_4/p_1, and on the physical properties of 'driver' and 'driven' gases. A diagram of the pressure and

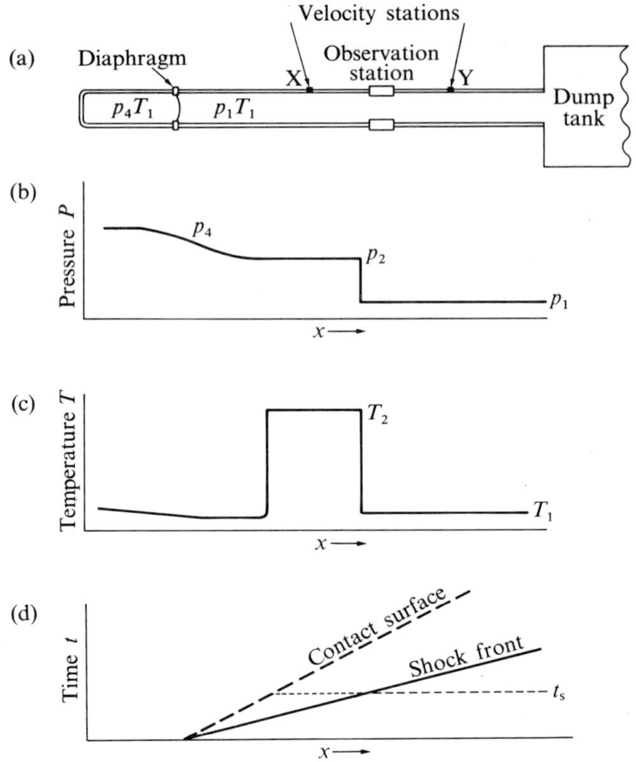

FIG. 2.9. The shock tube (showing pressure and temperature variation along the tube at time t_s).

temperature distribution along the tube after a shock front has been formed in a monatomic gas (constant C_V) is given in Figs. 2.9(b) and (c). Pressure and temperature rise sharply from p_1, T_1 to p_2, T_2 in the shock front and remain constant until the interface, or contact surface between driver and driven gas, is reached. Shock front and contact surface move at different speeds, as shown in Fig. 2.9(d). For a perfect monatomic gas, where C_V is independent of temperature, the values of p_2 and T_2 are given by

$$\frac{p_2}{p_1} = \frac{2\gamma M_{\mathrm{I}}^2 - (\gamma - 1)}{\gamma + 1} \tag{2.11}$$

$$\frac{T_2}{T_1} = \frac{\left(M_{\mathrm{I}}^2 - \frac{\gamma - 1}{2}\right)\left(\frac{\gamma - 1}{2} M_{\mathrm{I}}^2 + 1\right)}{\left(\frac{\gamma + 1}{2}\right)^2 M_{\mathrm{I}}^2} \tag{2.12}$$

For a real polyatomic gas, where C_V becomes temperature-dependent in the range T_1-T_2, more complicated treatment, involving the change of enthalpy with temperature, is needed. Detailed treatment of the theory of shock waves is given in standard works.[39,40,41]

2.3.2. Relaxation in a shock front

In a shock front the flow kinetic energy of the gas entering the front is transformed into random *translational* energy of the gas molecules. Since there is free interchange of translational energy at every collision (cf. § 1.2.2), equilibrium will be attained in only a few collisions, and the thickness of the shock front in a monatomic gas will correspond to a few mean free paths only, giving the almost vertical rise of T and p shown in Fig. 2.9. For most polyatomic gases *rotational* equilibrium is also attained in < 10 collisions, and rotational relaxation barely increases the thickness. *Vibrational* relaxation may, however, require several thousand collisions, and the temperature profile of the shock front will become modified as shown in Fig. 2.10(a). The density, ρ, of the gas

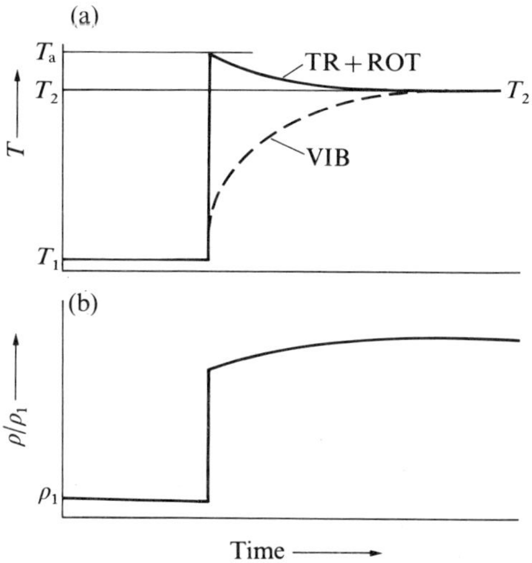

FIG. 2.10. Relaxation behind a shock front

behind the front is inversely proportional to T_{TR}, and will vary as shown in Fig. 2.10(b). If the density profile is accurately monitored as the shock front passes an observation point, the T-V relaxation time may be calculated, and may be assumed to refer to a temperature $(T_a - T_2)/2$. Two measurements are thus required:

(a) *The temperatures* T_2 and T_a, which may be calculated from the shock speed W_s and the Mach number M_I by the corrected form of eqn (2.12). The speed is measured by monitoring the passage of the shock past two 'velocity stations', X and Y in Fig. 2.9(a). The detectors at X and Y usually consist of piezo-electric pressure transducers, or of electrical temperature gauges depending on the conductivity of thin metal films.

(b) *The density profile*, which is measured through the windows of the 'observation station' (Fig. 2.9(a)) as the shock front passes. Since the refractive index of a gas varies with the density, this may be done by optical methods. The earliest method was the Mach-Zehnder interferometer shown in Fig. 2.11.

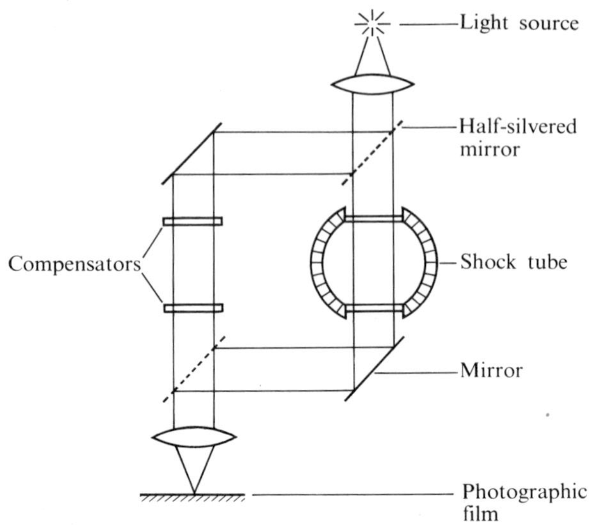

FIG. 2.11. Mach-Zehnder interferometer

The photographic record of the interference fringes produced by the shock front in CO_2 is shown in Fig. 2.12. Analysis of the curvature yields the relaxation time. A more accurate technique for small relaxation times is the laser-Schlieren method. This depends on the fact that a light beam is deflected by a refractive index gradient (as found in a shock front). The apparatus is shown schematically in Fig. 2.13. The detector signal can be interpreted to give an accurate density profile of the shock front as it passes. Examples of the records produced by non-relaxing and relaxing gases are shown in Fig. 2.14.

Other techniques, including the use of spectroscopic methods for monitoring the shock temperature, are described in the standard works already quoted. Shock tubes are suitable for measuring relaxation times ranging from 0·2 μs to 5 ms at temperatures in the range 500 to 5000 K, and most of the earlier data for vibrational relaxation of diatomic molecules have been obtained by this method.

2.4 Infrared excitation

2.4.1 Absorption and emission of radiation

If the matrix element of the dipole moment between two molecular energy states is non-zero, they may be transformed from one to the other by emission or absorption of radiation. Three basic processes are involved:

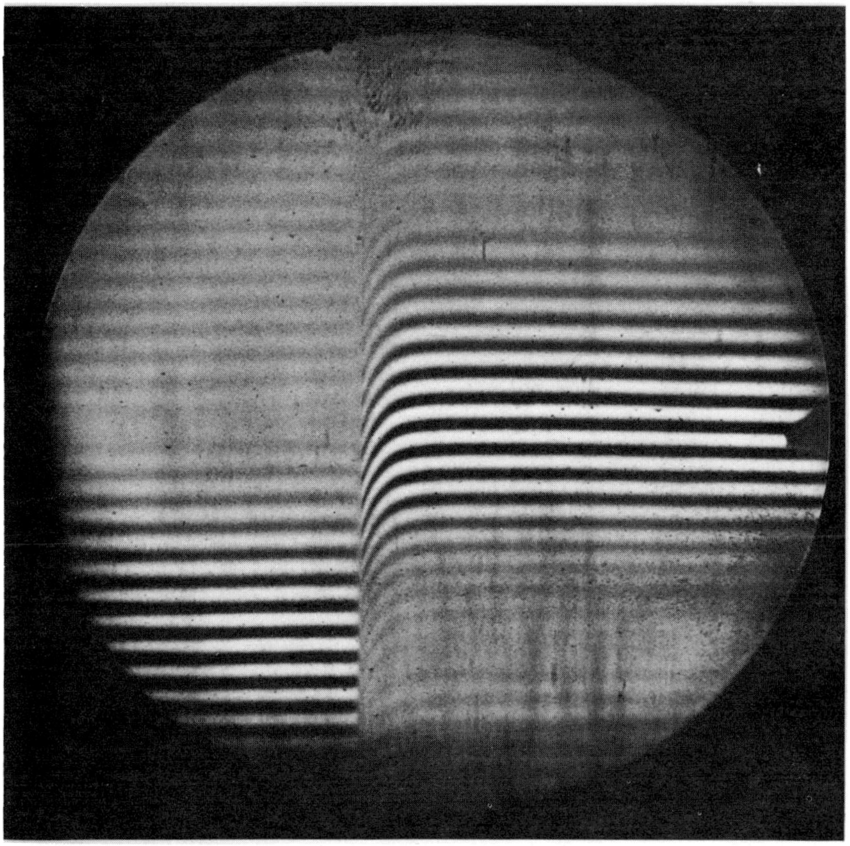

FIG. 2.12. Interferogram of a shock front in CO_2 (kindly supplied by Dr C.J.S.M. Simpson)

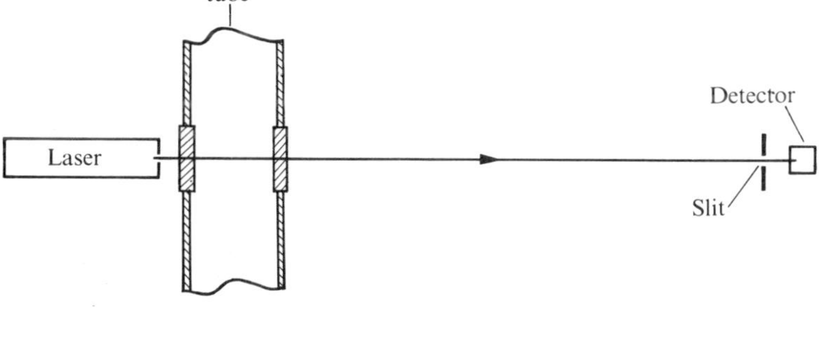

FIG. 2.13. Laser-Schlieren apparatus

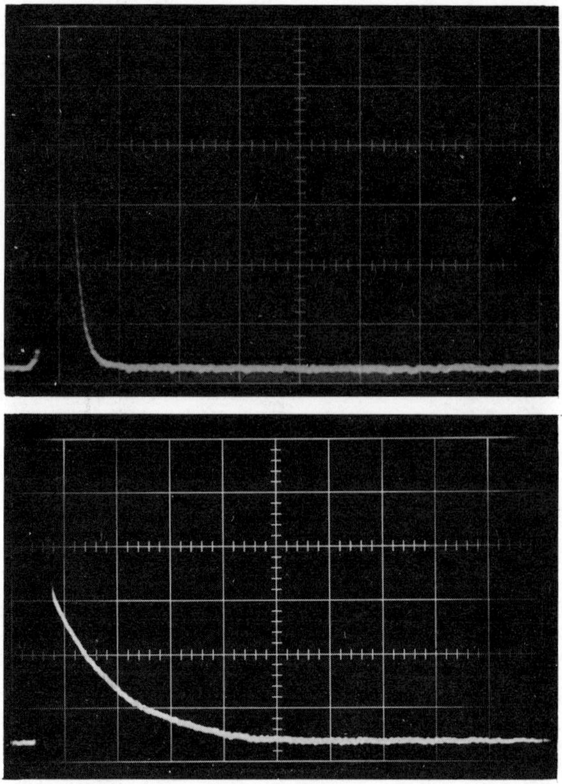

FIG. 2.14. Laser-Schlieren record of a shock-front. Top: non-relaxing gas. Bottom: relaxing gas. (Kindly supplied by Dr C.J.S.M. Simpson.)

(i) induced absorption; $M + h\nu \rightarrow M^*$
(ii) induced emission; $M^* + h\nu \rightarrow M + 2h\nu$
(iii) spontaneous emission; $M^* \rightarrow M + h\nu$.

If there are N_0 molecules in a (v=0) vibrational level and N_1 in a (v=1) level, the rate of change is given by

$$\frac{dN_1}{dt} = N_0 B_{01} U(\nu) - N_1 B_{10} U(\nu) - N_1 A_{10} \tag{2.13}$$

where $U(\nu)$ is the energy density of the radiation field at frequency ν, and B_{01}, B_{10}, and A_{10} are the Einstein coefficients for processes (i), (ii), and (iii) respectively. They are related by the equations

$$B_{01} = B_{10}$$

and

$$A_{10} = \frac{8\pi h\nu^3}{c^3} B_{10},$$

where c is the speed of light. If the system is irradiated for a short time, N_1 will be increased from its thermal equilibrium value. If the radiation is now switched off, N_1 will decay by process (iii) at a rate

$$\frac{dN_1}{dt} = -A_{10}N_1.$$

This is a relaxation process with a relaxation time, or radiative lifetime, $\tau = 1/A_{10}$, and will compete with the collisional relaxation process. (Induced emission by process (ii) due to thermal radiation will also occur, but may usually be neglected.)

2.4.2 Laser-excited vibrational fluorescence

Radiative lifetimes of vibrationally excited states lie in the range 1 to 100 ms. Collisional deactivation at pressures around 1 atm is usually several orders of magnitude faster; values of β range from 1 ns to 6 μs. Under these conditions β may be measured by observing the rate of decay of infrared fluorescence after activation. The intensity of fluorescence will give a direct measure of N_1, and will decay exponentially with time. The much slower radiative relaxation may be ignored.

The most appropriate source of radiation is a pulsed infrared laser. This is used to irradiate a sample tube containing the gas under investigation, and the decay of infrared fluorescence after the laser is shut off is followed by an infrared detector. A schematic apparatus for investigating vibrational relaxation of CO_2 (00^0 1) using a CO_2 laser as radiation source is shown in Fig. 2.15.

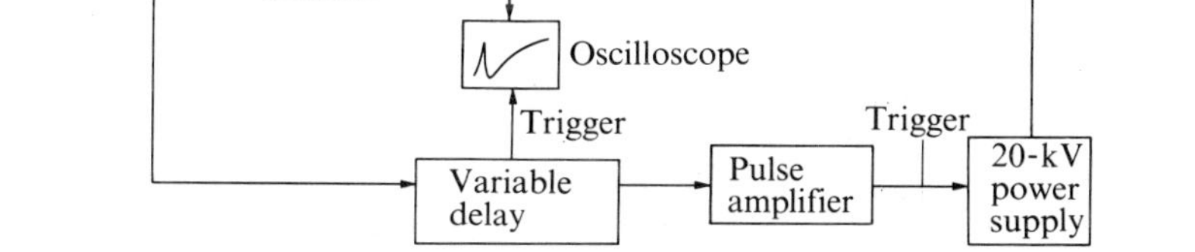

FIG. 2.15. CO_2 laser apparatus

The laser is Q-switched by a rotating mirror which forms one end of the optical cavity. A small light source and photo-detector focused on the opposite side of the mirror produce a signal which triggers both the laser power supply at the appropriate moment to produce an intense pulse of laser radiation when the optical cavity is aligned and the oscilloscope to record the decay of fluorescence after the pulse shuts off. The sample cell may be located either within the optical cavity (as shown in Fig. 2.15), or outside it.

A typical oscilloscope trace and semi-log plot of the exponential fall-off in intensity of fluorescence with time are shown in Fig. 2.16. Measurements can be made at varying pressures, but at lower pressures correction is necessary for deactivation of the activated species by diffusion to the cell wall. This method, which is very simple in principle, may be used for investigating the relaxation of any infrared active vibrational modes for which a suitable laser source and radiation detector can be found. The recent development of tunable infrared laser sources greatly increases its possibilities. Parametric oscillators, involving crystals of $LiNbO_3$, and tunable dye lasers[44] have been used.

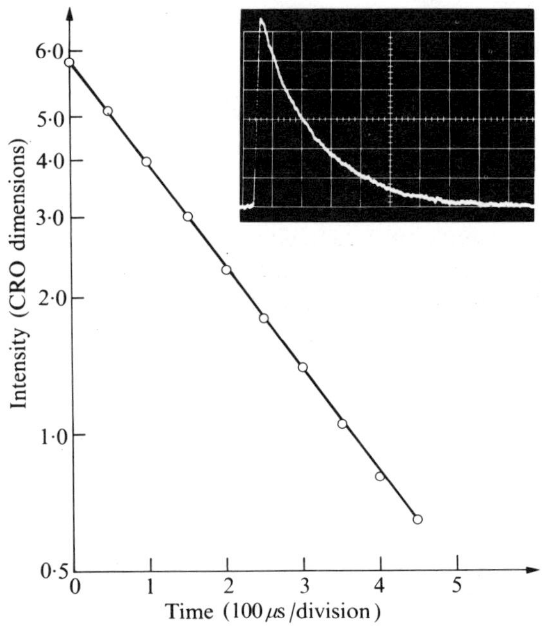

FIG. 2.16. Typical oscilloscope trace and semi-log plot in CO_2

The laser-fluorescence technique is an essential adjunct to the shock tube and acoustic methods for investigating low-level transitions in the vibrational modes of polar diatomic molecules at lower temperatures, where they are not thermally active. It is the only method for investigating relaxation of specified *upper* vibrational modes of polyatomic molecules, and is a good way of investigating V-V transfer between different molecules. Extensive reviews of laser-fluorescence work have been given by Bradley Moore.[45,46]

2.4.3. *Fluorescence quenching*

There is one diatomic gas, CO, which is almost non-polar, but shows strong infrared active vibrational transitions. The radiative lifetime of the v = 1-0 transition (2143 cm^{-1}) has been estimated as 33 ms.[47] The collision lifetime (owing to the absense of strong intermolecular forces) is of the same order as for non-polar diatomic gases, and has been estimated from shock-tube experiments to be greater than one second. This is *longer* than the radiative lifetime, in striking contrast to the polar molecules considered above, so that if pure CO is excited by radiation the whole the excitation energy is recovered as resonance fluorescence. Certain impurities, such as H_2, He, O_2, are highly effective energy transfer catalysts (cf. § 3.5.4), and can act as *fluorescence quenchers* by reducing the collision lifetime until it approach the radiative lifetime.

Millikan[47] developed a technique whereby a stream of very pure CO, surrounded by a 'jacket' of argon to eliminate wall collisions, was passed down the centre of a tube. The gas was excited by infrared radiation from a CH_4/O_2 flame, which raised the vibrational temperatures to >700 K while the rotational and translational temperatures were unaffected. The intensity of fluorescence was observed by an infrared spectrometer which was set at right-angles to the radiation source, and could be moved downstream. Small quantities of foreign gases were added, and the pressure of additive necessary to quench the fluorescence to half its original intensity was observed. Under these conditions the collision lifetime may be assumed to be exactly equal to the radiative lifetime (33 ms), and the collision efficiency for CO*/additive collisions can be calculated.

2.4.4. *Infrared and microwave double resonance*

An alternative to the fluorescence technique for monitoring the relaxation of an excited species formed by radiation is furnished by double-resonance spectroscopy. The Boltzmann distribution of population between two levels is disturbed by powerful monochromatic radiation of the exact frequency required, and the subsequent equilibration is followed by absorption spectroscopy at another frequency. The technique depends on the availability of powerful monochromatic sources of pumping radiation, and was first applied to the investigation of rotational transitions occuring in the microwave region, where powerful klystron sources are available. Oka[50] used steady-state pumping by a klystron source to saturate a microwave rotational transition in H_2CO, HCN, and $H_2C{:}CO$, and observed the change in absorption strength from a weaker klystron source operating at the frequency of a different 'signal' transition. Experiments using steady-state pumping give qualitative information about the collision-induced transitions between the rotational levels involved, but do not enable relaxation times to be observed. By using a modulated pumping system Unland and Flygare[393] were able to measure rotational relaxation times for COS. A theoretical study of the relation between modulated double resonance and inelastic collision processes has been published by Gordon.[394]

The development of infrared lasers as pumping sources enabled the technique to be extended to the infrared region. Infrared-microwave double resonance was used by Shimizu and Oka[59] to investigate rotational transfers in NH_3. Pulsed monochromatic infrared radiation from a N_2O laser produced a non-equilibrium rotational distribution within the v=0 level by pumping selected v=0, J, K, levels into v=1, J', K', levels. The relaxation of the rotational distribution in the v=0 state was then monitored by microwave absorption. Infrared double resonance was used by Steinfeld *et al.*[49] to follow the vibrational relaxation of SF_6 after excitation by an intense pulse from a Q-switched CO_2 laser by measuring the absorption of radiation from a much weaker c.w. CO_2 laser source. Ref. 49 gives a very full discussion of these complicated techniques, which were also used by Rhodes *et al.*[337] to investigate the relaxation of the $10^0\,0$ state of CO_2.

2.4.5. *Optic-acoustic effect–the spectrophone*

In 1881 Professor Tyndall reported on 'the action of an intermittent beam of radiant heat upon gaseous matter.'[53] His experiment is best described in his own words:

'My first source of rays was a Siemens' lamp connected with a dynamo-machine worked by a gas engine. A glass lens was used to concentrate the rays, and afterwards two lenses. By the first the rays were rendered parallel, while the second caused them to converge to a point about 7 inches distance from the lens. A circle of sheet zinc provided first with radial slits and afterwards with teeth and interspaces, cut through it, was mounted vertically on a whirling table, and caused to rotate rapidly across the beam near the focus. The passage of the slits produced the desired intermittence, while a flask containing the gas or vapour to be examined received the shocks of the beam immediately behind the rotating disk. From the flask a tube of india-rubber, ending in a tapering one of ivory or box wood led to the ear which was thus rendered keenly sensitive to any sound generated within the flask. . . . With this arrangement the number of sounding gases and vapours was rapidly increased. But I was soon made aware that the glass lenses withdrew from the beam its most effectual rays. The silvered mirrors employed in my previous researches were therefore invoked; and with them, acting sometimes singly and sometimes as conjugate mirrors, the curious and striking results which I now have the honour to submit to the [Royal] Society were obtained. Sulphuric ether, formic ether, and acetic ether being placed in bulbous flasks, their vapours were soon diffused in the air above the liquid. On placing these flasks, whose bottoms only were covered by the liquid, behind the rotating disk, so that the intermittent beam passed through the vapour, loud musical tones were in each case obtained.'

Other gases found to produce loud musical tones were N_2O, CO_2, NH_3 and *moist* air (saturated with H_2O): all good infrared absorbers.

This clearly demonstrated that the absorption of periodically chopped infrared radiation by an absorbing gas results in the production of sound. The periodic vibrational excitation produced by the radiation is converted by

V-T transfer into a periodic fluctuation of translational temperature and pressure, which is a sound wave. The rate of conversion depends on the vibrational relaxation time: this will influence both the *phase-shift* due to time-lag between radiative and sonic pulses, and the variation of *amplitude* of the sound waves with chopping frequency. Both of these effects can be investigated with a *spectrophone* which consists essentially of a small gas cell with a rock salt window to admit chopped infrared radiation, and a condenser-transducer (cf. § 2.2.2) to detect the sound. Attempts to use the spectrophone for measuring relaxation times started in 1948,[54] and have been recently reviewed by Read.[55] There have been formidable difficulties, both experimental and theoretical. Experimentally great trouble has been caused by gas contamination, apparently due to desorption of impurities from the cell wall under the influence of the radiation. Theoretically the actual interpretation of the measurements is extremely complicated: for example, a rise in pressure of the absorbing gas, which would be expected to decrease the relaxation time and the phase-shift, may in fact cause an apparent increase in the measured phase-shift due to all the energy being absorbed closer to the cell window and further from the transducer.[56] These difficulties *can* be overcome, and a careful review of earlier results by Huetz-Aubert *et al.*[57] shows that they can be brought into agreement with results of other techniques by suitable treatment. But the spectrophone has been effectively superceded by the newer infrared techniques described above. It is particularly unsuited for investigating relaxation of upper vibrational modes in polyatomic molecules. Cottrell *et al.*,[58] who have been responsible for much of the published work on spectrophones, remark: 'What is observed in the spectrophone is not the decay of vibrational energy from a specific mode, but the appearance of translational energy in a gas in which the molecules have been excited in a specific mode. This means that the "relaxation time" observed in the experiment is not the average lifetime of the mode excited, but a measure of the time for which the energy, initially put into one mode, remains locked up, probably in a variety of vibrational modes. Thus the overall "relaxation time" observed may well be a complicated function of a number of different relaxation times for precisely defined processes.'

2.5. Raman excitation

Vibrational transitions which are not infrared active, but involve a change in polarizability, may be excited by the Raman effect. This is illustrated on an energy-level diagram in Fig. 2.17(a). If an incident photon of visible light, frequency ν, collides with a molecule, it can be scattered inelastically after losing part of its energy in exciting the molecule by an energy quantum, ΔE, corresponding to the energy separation of the $v = 0$ and $v = 1$ vibrational levels. The scattered radiation will have a lower frequency $\nu - \Delta\nu$, where $\Delta E = h\Delta g$. This is normally a weak effect: the majority of the incident photons undergo elastic scattering without change of frequency; the inelastically scattered radiation is weak and only a small number of molecules are raised into the

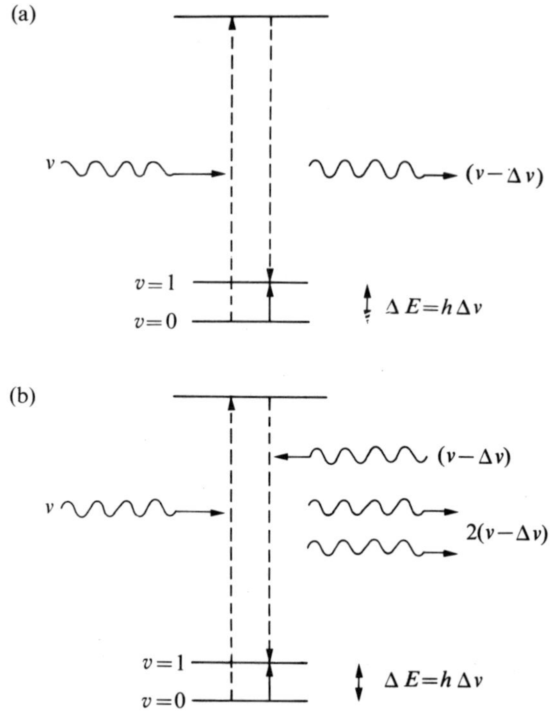

FIG. 2.17. (a) Raman excitation (Stokes). (b) Stimulated Raman excitation

$v = 1$ level. A much more intense stimulated Raman excitation may be produced by exciting the molecule with two incident light waves of frequency differing by exactly $\Delta\nu$. (cf. § 2.4.1, process (ii)) This is shown in Fig. 2.17(b): it has the advantage that the two light sources may be tuned to correspond exactly with a particular molecular transition, and this will be selectively excited. If pulsed laser sources are used, the relaxation of the excited level can be followed after the pulse has shut off.

The apparatus shown schematically in Fig. 2.18 has been used by Ducuing *et al.*[51] for measuring vibrational relaxation of N_2 and O_2 at room temperature. A pulsed ruby laser and a tunable dye laser are aligned at opposite ends of a gas cell, filled to a pressure between 15 and 60 atm. The ruby laser pumps the dye laser, which is tuned so that $\nu_{\text{ruby}} - \nu_{\text{dye}} = \Delta\nu$ for vibrational excitation. The two beams are focused together into a small focal zone in the centre of the gas cell where intense vibrational excitation occurs. V-T relaxation then produces changes in translational temperature, density, and refractive index which vary radially in the focal zone. This gives rise to a time-dependent 'lensing' effect, which may be monitored by passing the beam from a He-Ne laser, with appropriate screening devices, perpendicularly through the focal zone on to a photomultiplier. The signal from the photomultiplier will show an exponential decay, resembling that shown in Fig. 2.16, and the relaxation time

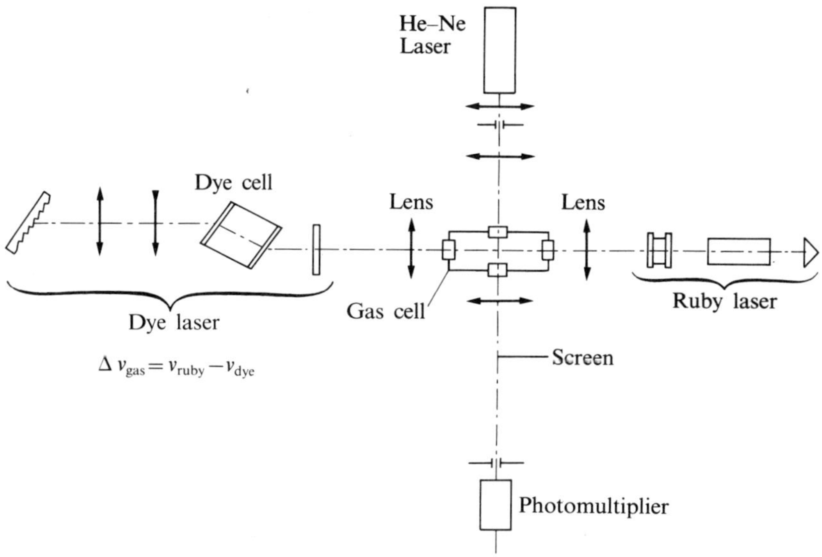

FIG. 2.18. Apparatus for Raman excitation[51]

can be calculated from the logarithmic plot. This method can be used, in principle, for any Raman-active transition, and forms an invaluable extension to infrared laser techniques for investigating transitions in non-polar molecules such as the homonuclear diatomics in the low temperature range, where the vibrational mode is not thermally active. A less sophisticated stimulated Raman technique was used to determine the vibrational relaxation time of H_2, in which the lower frequency radiation was supplied by the Stokes-scattered photons themselves when gas at a high pressure was irradiated by a very powerful *Q*-switched ruby laser.[52]

2.6. Electronic excitation

2.6.1. *Fluorescence*

Absorption of radiation in the ultraviolet region can raise molecules into an upper electronic state, with associated changes in vibrational and rotational levels. A schematic energy-level diagram showing the vibrational changes is given in Fig. 2.19. Absorption of monochromatic ultraviolet radiation raises the molecule from the $v''=0$ level of the electronic ground state A to the $v'=3$ level in the upper electronic state B. The excited molecule can now revert to a variety of v'' levels of state A by spontaneous ultraviolet emission. The rates of absorption and emission of radiation are governed by the principles discussed in § 2.4.1: ultraviolet frequencies are much higher than infrared, and the radiative lifetime of the excited state is much shorter, about 1-100 ns, which is of the same order as rotational and vibrational collisional relaxation times at atmospheric pressure. The pure, spontaneous emission spectrum can be

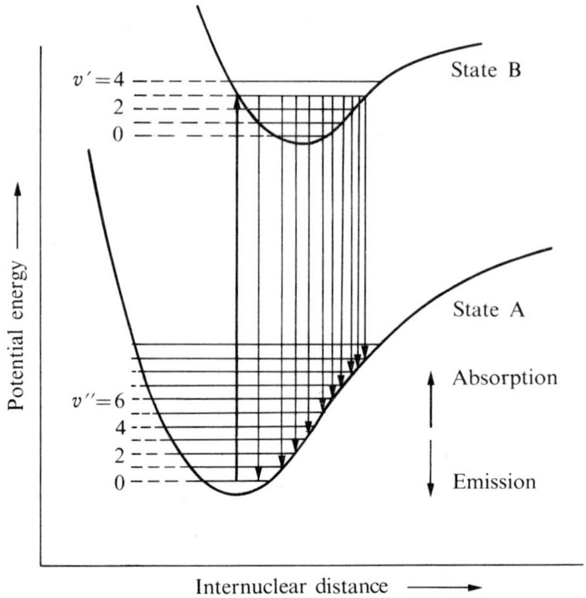

FIG. 2.19. Electronic fluorescence

observed at pressures < 2 torr, where the gas kinetic collision rate falls to <0·1 ns^{-1} (see Table 1.1), so that the majority of excited molecules undergo radiative decay before suffering a collision. As the pressure is raised, the collision rate increases, and collisional vibration and rotational relaxation in the B state occurs before the electronic excitation decays to the A state by fluorescence. This modifies the fluorescence spectrum (it may in addition cause electronic quenching), and observation of the relative intensities of fluorescence lines at varying pressures, or in the presence of additives, yields detailed information about vibrational and rotational relaxation times in the B state. The application of this method to energy transfer in H_2 ($B'\,\Sigma_u^+$) is described in detail by Bradley Moore *et al.*,[60] who give references to earlier work.

2.6.2. *Flash photolysis*

The technique of flash photolysis, developed by Norrish for investigation of photochemical processes, has been used for determining vibrational relaxation times of diatomic molecules in the microsecond region. If, for example NO or CO are irradiated by a powerful ultraviolet flash,[61,62] a substantial population of molecules is excited into an upper electronic state (cf. state B in Fig. 2.19). Rapid radiative decay occurs, as described above, to various v'' levels in the lower electronic state. This is followed by rapid relaxation of upper v'' levels to the $v''=1$ level by V-V transfer. A substantial population of molecules in the $v''=1$ level is produced, and its relaxation to $v''=0$ by the much slower V-T

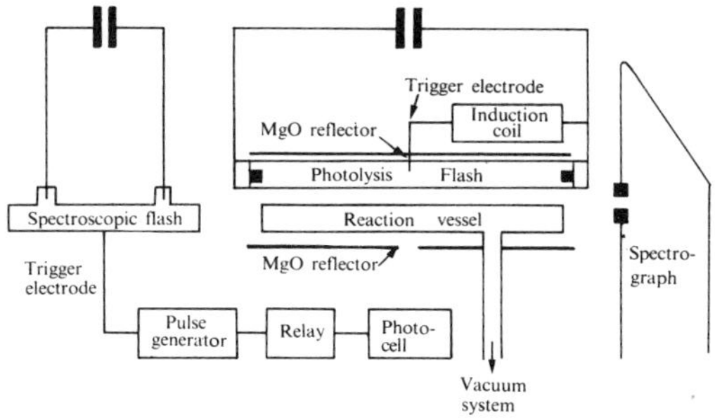

FIG. 2.20. Schematic diagram of a flash photolysis apparatus

process may be followed by kinetic absorption spectroscopy. The apparatus is shown diagrammatically in Fig. 2.20. The gas cell is a quartz tube, 2 cm diameter and about 20 cm long. A 'photolysis flash' tube placed parallel to this produces the excitation. A weak 'spectroscopic flash', which can be triggered to occur at varying intervals after the photolysis flash, enables the absorption spectrum of the gas to be recorded on a photographic plate. A typical photographic record for NO is shown in Fig. 2.21. The decay of the 0, 1 band at 2370 Å is clearly

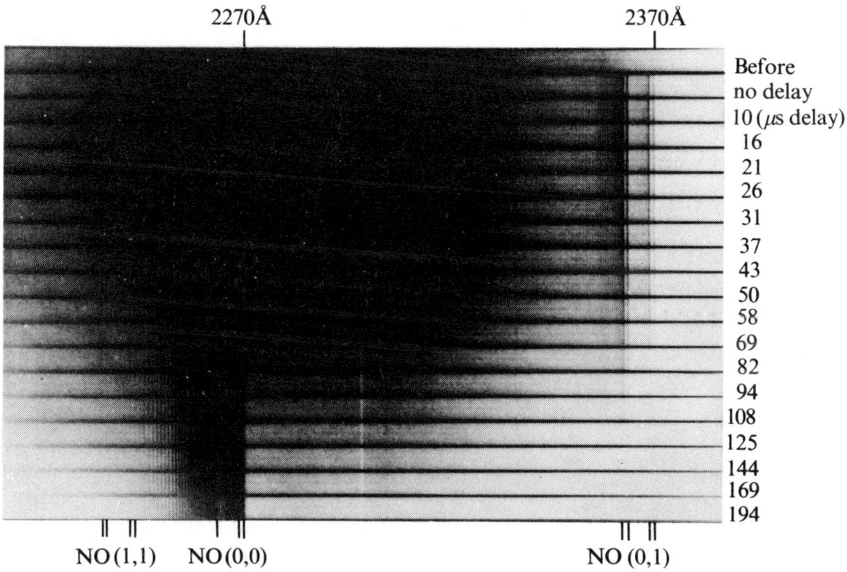

FIG. 2.21. Photographic record of decay of NO $^2\Pi$ (v=1) (kindly supplied by Dr A.B. Callear)

visible: the relaxation rate of the $v''=1$ level may be calculated from measurements of the plate density with a microdensitometer. A modified apparatus, using a microwave flash to produce $NO^2\ \Pi$ ($\nu=1$) has been used by Callear and Williams[63] to investigate the effect of additives which would be decomposed by the photo-flash.

The flash photolysis apparatus can also be used to study relaxation of vibrationally active species produced by a photochemical *reaction.* The flash photolysis of HBr to give H and Br atoms, in the presence of excess Br_2, gives a reaction[64]

$$H(1^2\,S_{\frac{1}{2}}) + Br_2 \rightarrow HBr(v''=n) + Br(4^2\,P_{\frac{1}{2}}).$$

HBr (v''=n) is rapidly deactivated to $v''=1$ by V-V transfer, and the V-T relaxation of HBr ($v''=1$) may be followed as described above.

2.7. Chemical excitation–infrared chemiluminescence

In an exothermic reaction between an atom and a diatomic molecule:

$$A + BC \rightarrow AB^* + C; \quad \Delta H = -E,$$

a large proportion of the heat of reaction E may appear as vibrational and/or rotational excitation of the product molecule, AB*. If AB* is polar, this leads to emission of infrared radiation as *chemiluminescence.* Since radiative lifetimes in the infrared are much longer than collision lifetimes (see § 2.4.2), the rate of collisional deactivation of AB* may be followed by monitoring the decay of infrared luminescence. The type of apparatus used by Polanyi *et al.*[65] for investigating vibrational relaxation of HCl*, produced by the reaction

$$H + Cl_2 \rightarrow HCl^*\,(v=1 \text{ to } 5) + Cl,$$

is shown in Fig. 2.22. It consists of a 11·4 cm diameter pyrex tube with gas inlet tubes and with three sapphire windows, spaced at 5·7 cm intervals along the line of flow of the gas. The interior of the tube is gold-plated in the neighbourhood of the windows so as to produce maximum reflection of radiation. H atoms (produced by a Wood's discharge tube) and Cl_2 gas are fed into the top end at measured flow-rates and pressures, and the emitted radiation at different distances downstream is monitored by an infrared grating spectrometer. By observing the effects of varying pressures, flow-rates, and of addition of additives, relaxation times may be estimated. This is the best method available for investigating relaxation of *higher* vibrational levels in diatomic molecules. Vibrational relaxation of CO in levels up to $v=13$, produced by the reactions

$$O(^3P) + CS_2 \rightarrow SO + CS$$
$$O(^3P) + CS \rightarrow CO(4 \leqslant v \leqslant 13) + S; \ \Delta H = -75 \text{ kcal mol}^{-1}$$

has been investigated by Hancock and Smith.[66] The method can also be applied to rotational relaxation from higher J-levels, whose collision lifetimes are much

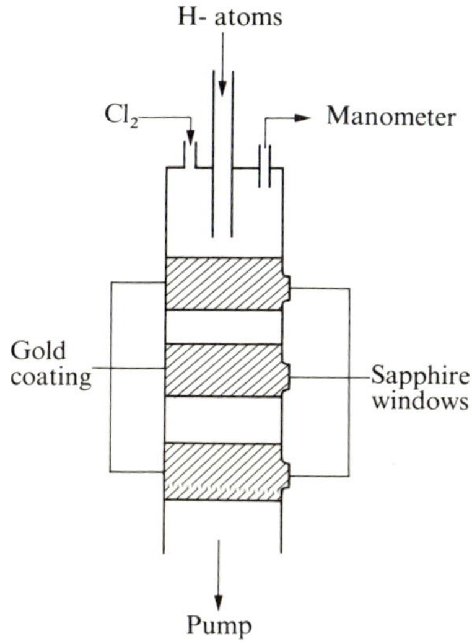

FIG. 2.22. Infrared chemiluminescence apparatus

longer than for lower levels (cf. § 1.3.2): relaxation of HCl (J=13) has been studied by Polanyi and Woodall.[67]

2.8. Molecular beams

The basic process investigated by molecular beams is the inelastic collision between an atom (or ion) A and a diatomic molecule BC. The conditions are chosen so that no chemical reaction occurs, but there are changes in the vibrational and/or rotational states of BC leading to compensating changes in the velocity of both partners. A beam of BC molecules in selected velocity and internal energy states is crossed at a determined angle with a beam of A atoms (or ions) of selected velocity. Collisional scattering occurs, and the angular distribution, velocity, and internal energy state after the collision are monitored by a detecting system. A schematic diagram of the experiment is shown in Fig. 2.23. The entire apparatus is in a vacuum ($\sim 10^{-6}$ torr) such that the mean free path is much greater than the dimensions of the apparatus, and collisions only occur in the scattering region where the two beams cross.

The detector measures, in principle, the number of molecules in an internal energy state j scattered into an element of solid angle $d\omega$. If the internal energy state of the molecules in the incident beam is i, a *differential cross-section* for the process $i \rightarrow j$ may be defined as:

$$\sigma_{ij}(\omega)\,d\omega = \frac{\text{No. of molecules in state } j \text{ scattered per second into } d\omega}{\text{Intensity of incident molecules in state } i}.$$

If this is integrated over all angles, a *total cross-section* σ_{ij} is obtained for the process. Since this is a single process involving only two internal states, x_{ij} is identical with the collision cross-section Q defined in § 1.4.2, so that

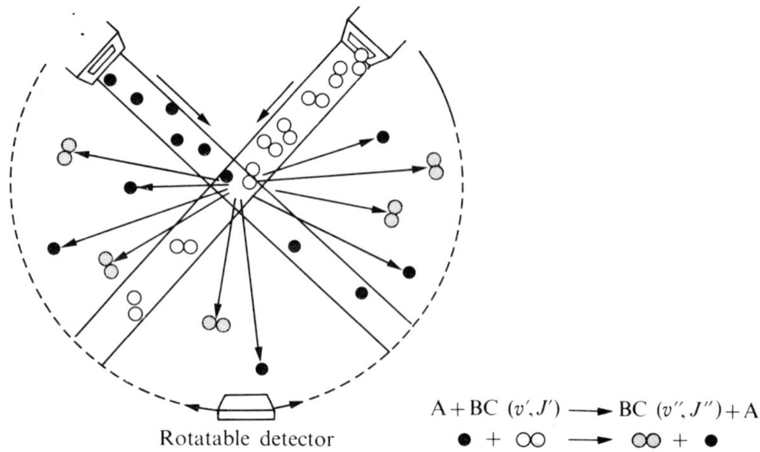

FIG. 2.23. Schematic diagram of a molecular-beam scattering experiment[70]

$$\sigma_{ij}/\pi\sigma^2 = P_{ij},$$

where $\pi\sigma^2$ is the gas kinetic cross-section, and P_{ij} the probability of energy transfer per collision. The outstanding advantage of molecular beam studies of energy transfer is that, in addition to the total cross-section for the process, they give information about the angle-dependence of the differential cross-section. This will depend critically on the actual form of the intermolecular potential energy surface; varying angles of deflection correspond to different impact parameters, and this enables inferences to be made about the relative efficiencies of the repulsive and attractive parts of the potential (cf. § 1.2.2). For very simple systems exact calculations may be made for comparison with experiment.

Highly sophisticated apparatus is needed for molecular beam studies of molecular energy transfer, since they involve monitoring the velocity and/or internal energy states of the participants, in addition to the simple measurement of angular distribution of intensity. The majority of experiments reported use a technique for monitoring the *velocities* before and after collision, from which the *energy-change* involved in the collision may be inferred. Velocity selectors, which consist of systems of rotating slotted disks, can be used both in producing the incident beam and in making a velocity analysis of the scattered beam on its path to the detector. Where beams of charged ions are concerned, electrostatic velocity selection may be used. A time-of-flight method, employing variable path length between scattering centre and detector, has also been widely used for energy-change measurements. A schematic diagram of a time-of-flight apparatus used by Toennies[70] for investigating collisions of Li^+ and H_2 is shown in Fig. 2.24. Here the velocity of the *incident* Li^+ beam is electrostatically selected. An alternative technique, which is more difficult and has been less widely used, is to monitor the *internal* energy states of the scattered molecules.

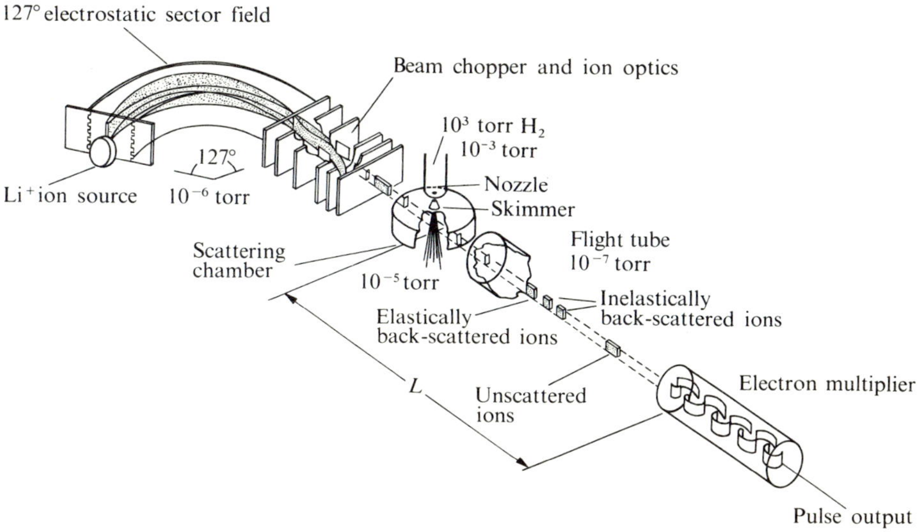

FIG. 2.24. Schematic diagram of time-of-flight apparatus[70] for studying inelastic scattering in Li^+-H_2. The Li^+ beam source is at the left. The ion beam energy is selected by a 127° electrostatic sector field. The ion beam is then electrically chopped and passed into the scattering chamber, where it is crossed with an H_2 nozzle beam. The flight path length L is variable between 50 and 170 cm.

Electron resonance analysis, by passing the scattered beam into an electron resonance spectrometer, has been used for monitoring the vibrational states of LiF scattered by various atoms and molecules.[71] Specific rotational states of TlF have been separated by an electrostatic quadrupole field.[72] Infrared spectroscopy[73] has been used for monitoring rotational states of HCl, and electron beam fluorescence spectroscopy[74] for N_2. Progress in the whole field was reviewed by Toennies[70] in 1974: it is limited in its application to simple molecules, atoms, and ions for which suitable detectors and energy selectors or analysers can be devised.

2.9. Transport properties

According to elementary kinetic theory, the viscosity η and thermal conductivity k of a gas are related by the equation

$$k = \eta C_V/M, \quad \text{or} \quad kM/\eta = C_V.$$

More sophisticated theory[75] shows that for smooth spherically symmetrical molecules,

$$kM/\eta = 2{\cdot}5\, C_V$$

The increase in the actual ratio k/η is mainly due to the persistance of forward velocity in collisions. The derivation of this equation assumes completely free energy transfer in every collision, and it is closely followed by the monatomic gases, where only translational energy is involved. Polyatomic gases, possessing vibrational and rotational energy which are not freely exchanged with translation in collision, show experimental values of the ratio kM/η which are systematically lower than 2·5 C_V. Eucken[76] put forward the assumption that vibrational and rotational relaxation are so slow as to be negligible in this context, so that the contributions of translational and internal energy to the thermal conductivity can be regarded as independent. This gives rise to

$$kM/\eta = f_{\text{trans}}\, C_{\text{trans}} + f_{\text{int}}\, C_{\text{int}}\,. \tag{2.15}$$

The translational factor f_{trans} will be 2·5 for simple non-polar molecules. Transport of internal energy will be a simple diffusion process, and $f_{\text{int}} = \rho D/\eta$, where ρ is the density and D the diffusion coefficient. f_{int} takes the value ~1·3 for many molecules. Eqn 2.15 fits reasonably well for non-polar molecules at temperatures in the region of 300 K, and may be expressed in terms of an *Eucken Factor*, f_{E}, derived from (2.15) on the assumptions that $f_{\text{trans}} = 5/2$ and $C_{\text{trans}} = 3/2\,R$:

$$f_{\text{E}} = kM/\eta\, C_V = \frac{\rho D}{\eta} + \frac{3}{2}\left(\frac{5}{2} - \frac{\rho D}{\rho}\right)\frac{R}{C_V}\,. \tag{2.16}$$

Many polyatomic molecules and all polar molecules give experimental values of $kM/\eta C_V$ consistently lower than predicted by eqn (2.16). Mason and Monchick[22] have tried to explain this quantitatively on the assumption that, in these cases, comparatively rapid energy transfer between translational and internal modes renders the Eucken assumption invalid. Starting from the semi-classical kinetic theory of Wang Chang and Uhlenbeck, making simplifying assumptions, and taking into account collision numbers Z_{rot} and Z_{vib} for internal energy transfer, they derive an alternative function:

$$f_{\text{MM}} = kM/\eta C_V = f_{\text{E}} - \frac{2}{\pi C_V}\left(\frac{5}{2} - \frac{\rho D}{\eta}\right)^2 \left[\frac{C_{\text{rot}}}{Z_{\text{rot}}} + \frac{C_{\text{vib}}}{Z_{\text{vib}}}\right] \tag{2.17}$$

It is at once clear that only very rapid relaxation, leading to small values of Z_{rot} and Z_{vib}, would be expected to produce a significant effect, which means that, for most molecules, only rotational relaxation comes into consideration.

While the Mason-Monchick theory throws interesting light on the physics of the situation, it does not provide reliable values of Z_{rot} or Z_{vib}. On the experimental side, accurate measurements of k and η are not easily made, and small changes in the ratio k/η have a large effect on collision numbers calculated from eqn (2.17). On the theoretical side, Cowling[77] has pointed out that the simplifying assumptions necessary in this theory impose severe limits on its quantitative accuracy. Published values of Z_{rot} obtained by this method form transport measurements for a number of simple polar and non-polar molecules are of the correct order of magnitude, but consistently lower than values obtained from acoustic measurements. This may be due to the fact that acoustic data refer only to R-T transfer, whereas resonant R-R transfer between molecules of the same species must also play an important role in thermal conductivity, and is shown in § 5.4 to be highly efficient.

3

VIBRATION–TRANSLATION TRANSFER

3.1. The SSH theory

3.1.1. Introduction

The elementary discussion of inelastic collisions in Chapter 1 will have shown the difficulties in the way of any exact and rigorous theory of energy transfer. Even if a completely satisfactory theory were available, its quantitative application would be severely limited by the lack of accurate and realistic intermolecular potentials. The theory which has been most widely used for quantitative comparison with experimental measurements of vibrational relaxation is the Schwartz, Slawsky, and Herzfeld (SSH) theory. This accounts semi-quantitatively for most of the principal features of vibrational energy transfer, and gives fair quantitative predictions for simple molecules. We shall use it to provide a basic theoretical framework, and to illustrate the relative importance of the various factors which influence transfer probabilities. It is a quantum-mechanical theory, and treats the collision as a parallel stream of molecules, which may be regarded as a de Broglie wave, falling on a molecule regarded as stationary. The problem is handled by the distorted wave approximation (DWA) due to Jackson and Mott,[78] simplified by the modified wave-number approximation (MWN) due to Takayanagi.[79] The theory was first developed as a one-dimensional treatment by Schwartz, Slawsky, and Herzfeld[80] in 1952, extended as a three-dimensional treatment for diatomic molecules by Schwartz and Herzfeld[81] in 1954, and further extended to polyatomic molecules by Tanczos[27] in 1956. A full discussion is given by Herzfeld and Litovitz.[5]

3.1.2. Intermolecular potentials

The first requirement in any theory of energy transfer is to define an intermolecular potential $V(r)$. (cf. § 1.2.1) Analytical solutions of the collision problem require an *exponential* potential, as was used in the classical Landau-Teller theory (§ 1.4.1). The potential used by SSH theory is

$$V(r) = U\exp(-\alpha r) - \epsilon. \qquad (3.1)$$

This is shown diagrammatically in Fig. 3.1, where it is compared with the Lennard-Jones potential (Fig. 1.1.). It is essentially a repulsive potential, and the attractive forces, represented by $-\epsilon$, are regarded as simply increasing the velocity of relative motion, thereby favouring a higher transition probability. Molecular collision parameters, as shown in Table 1.1, are derived from experimental measurements of transport properties and are mostly expressed in

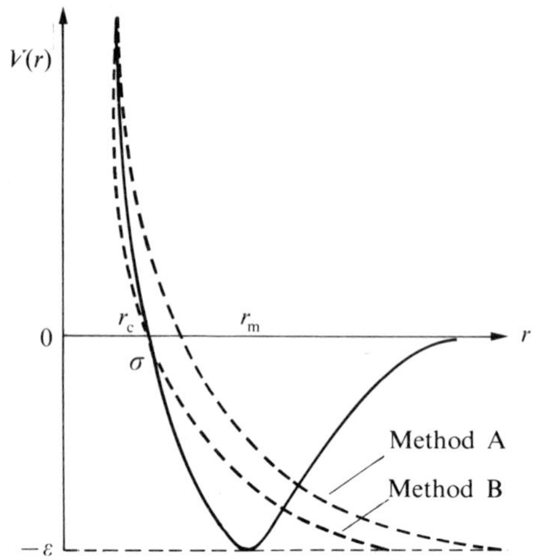

FIG. 3.1. 'Fitting' of exponential potential

terms of the Lennard-Jones potential: they have to be 'fitted' to eqn (3.1). There are two methods for doing this,[5] which are illustrated in Fig. 3.1. Both methods equate the two potentials at the classical turning point, r_c. Method A makes the exponential curve tangential to the Lennard-Jones curve at the classical turning point. Method B also equates the two curves at a second point, σ. Since neither the Lennard-Jones nor the exponential curve is more than an approximation to reality, the whole procedure may seem artificial, but some value of α has to be found. Method B, which gives a larger value of α, is usually found to agree better with experiment. Both r_c and α will vary with the velocity of approach, and the value α^* corresponding to the molecular speed with the highest transition probability, v^*, is used in the calculations (cf. Fig. 1.6). Similar 'fitting' techniques may be applied to potentials other than the Lennard-Jones.

3.1.3. Transition probabilities

The probability $P_{k-l(\text{b})}^{i-j(\text{a})}$ that during a binary collision vibrational mode (a) of one molecule will change its quantum state from i to j, while simultaneously a second mode (b) in the other molecule will change its state from k to l, is given by SSH theory as:

$$P_{k-l(\text{b})}^{i-j(\text{a})} = P_0(\text{a})\,P_0(\text{b})\,[V^{ij}(\text{a})]^2\,[V^{kl}(\text{b})]^2 \times$$

$$\times\,(4\mu/kT)\exp[-\epsilon/kT]\left(\frac{8\pi^3\mu\Delta E}{h^2}\right)^2\int_0^\infty f(\bar{u})\,\mathrm{d}\bar{u} \qquad (3.2)$$

where

$$f(\bar{u}) = \frac{\bar{u}}{\alpha^4}\cdot\left(\frac{r_c}{\sigma}\right)^2\exp\left[-\frac{\mu\bar{u}^2}{2kT}\right]\left\{\frac{\exp[L-L']}{(1-\exp[L-L'])^2}\right\},$$

$$L = \frac{4\pi^2\mu}{\alpha h}\cdot\bar{u} \quad \text{and} \quad L' = \frac{4\pi^2\mu}{\alpha h}\cdot\bar{v},$$

μ is the reduced mass of the colliding pair which approach one another with an effective relative velocity $\bar{\mathbf{u}}$ and recede with velocity $\bar{\mathbf{v}}$. The integration is over the thermal distribution at T of the molecular velocities assumed to be Maxwellian. ΔE is the amount of energy exchanged between the vibrational and translational degrees of freedom:

$$-\Delta E = h\nu_a(i-j) + h\nu_b(k-l) = (\bar{\mathbf{v}}^2 - \bar{\mathbf{u}}^2)\mu/2.$$

In a simple V-T transition, only mode (a) changes state, and $k = 1$ so that the pre-exponential term $[V^{kl}\,(\mathrm{b})]^2$ becomes unity (see eqn (3.4) below). SSH calculations are best performed by making a numerical integration of eqn (3.2).

When ΔE is large the integration can be solved analytically in terms of α^* to give the formula deduced by Tanczos,[27] which is valid for values of $\Delta E > 200$ cm^{-1} and forms a more convenient basis for discussing the influence of different factors on the transition probability. The Tanczos equation is

$$P_{k-l(\mathrm{b})}^{i-j(\mathrm{a})} = P_0(\mathrm{a})\,P_0(\mathrm{b})\left(\frac{r_c^*}{\sigma}\right)^2 [V^{ij}(\mathrm{a})]^2\,[V^{kl}(\mathrm{b})]^2\; 8\left(\frac{\pi}{3}\right)^{\frac{1}{2}} \times$$

$$\times\left[\frac{8\pi^3\mu\Delta E}{\alpha^{*2}h^2}\right]^2 \zeta^{\frac{1}{2}} \exp\left[-3\zeta + \frac{\Delta E}{2kT} + \frac{\epsilon}{kT}\right] \tag{3.3}$$

where

$$\zeta = \frac{\mu\mathbf{u}^{*2}}{2kT} = \left(\frac{\Delta E^2\mu\pi^2}{2\alpha^{*2}\hbar^2kT}\right)^{\frac{1}{3}}.$$

The Tanczos equation can be considered in four parts: (a) the geometrical, or steric, factors P_0; (b) the collision cross-reference factor $(r_c^*/\sigma)^2$, (c) the vibrational factors $[V]^2$; (d) the translational factor represented by the remainder of the equation.

(a) *The steric factors*, P_0, are required to account for the fact that the molecules are not spherically symmetrical and that some *collision orientations* will be more effective than others. They are usually taken as $\frac{1}{3}$ (the average of $\cos^2\theta$ taken over a sphere) for diatomic molecules and for longitudinal vibrations of linear polyatomic molecules. For non-linear polyatomic molecules and for bending modes of linear molecules, P_0 is taken as $\frac{2}{3}$. The theory takes no account of rotational transitions, and assumes the orientation to remain constant during a collision. These assumptions are invalid for hydrogen and hydrides where a very low moment of inertia is associated with rapid rotation and with wide spacing of rotational quantum levels; further discussion of these effects will be given later. The fact that certain orientations will be preferred for collisions involving polar molecules (cf. § 1.6) is also ignored, which limits the accuracy of the treatment for these molecules at lower temperatures.

(b) *The collision cross-reference factor*, $(r_c^*/\sigma)^2$, is the ratio of the intermolecular separation at the classical turning point to the separation at zero energy, σ, which is used in calculating the gas kinetic collision number Z (§ 1.2.3). It converts probabilities calculated for a collision cross-section πr_c^2, to the usual elastic cross-section of kinetic theory, $\pi\sigma^2$.

(c) *The vibrational factor,* $[V^{ij}\,(\mathrm{a})]^2$, is the square of the matrix element for the transition between the two vibrational states i, j of molecule (a). It represents the coupling of the initial and final states, i and j, of the oscillator under the influence of the perturbation produced by the collision, and corresponds to the square of the Fourier component of the varying force at the oscillator frequency which is found in classical theory (§ 1.4.1). It depends on the repulsion parameter α and the frequency and amplitude of the vibration. Vibrational factors are defined for zero, single, and double quantum jumps as:

$$[V^{i\text{-}i}]^2 = 1 \tag{3.4}$$

$$[V^{(i+1)\text{-}i}]^2 = [V^{i-(i+1)}]^2 = \alpha^2(\overline{A^2})\,(i+1)/2\gamma \tag{3.5}$$

$$[V^{(i+2)\text{-}i}]^2 = [V^{i-(i+2)}]^2 = \alpha^4(\overline{A^4})\,(i+1)\,(i+2)/16\gamma^2 \tag{3.6}$$

where $\gamma = 4\,\pi^2\,\nu/h$. The vibrational amplitude coefficient (A^2) is the Cartesian displacement of an atom for unit change of the normal coordinate of a given vibration. The values of (A^2) for the surface atoms of a molecule, which is treated as a 'breathing sphere', are then averaged and summed over the number of surface atoms to give $(\overline{A^2})$. For a diatomic molecule XY the amplitude coefficient depends simply on the masses of the atoms X and Y:

$$(\overline{A^2}) = A_{\mathrm{X}}^2 + A_{\mathrm{Y}}^2 = \frac{m_{\mathrm{Y}}}{m_{\mathrm{X}}(m_{\mathrm{X}}+m_{\mathrm{Y}})} + \frac{m_{\mathrm{X}}}{m_{\mathrm{Y}}(m_{\mathrm{Y}}+m_{\mathrm{X}})}.$$

For polyatomic molecules $(\overline{A^2})$ must be obtained from a normal coordinate analysis of the molecular vibrations.[82] Some values of $(\overline{A^2})$ for the vibrational modes of representative diatomic and polyatomic molecules are listed in Table 3.1: they reveal the striking influence of the presence of hydrogen and deuterium atoms due to their low relative mass. For the diatomic and triatomic molecules the values of $(\overline{A^2})$ are an order of magnitude larger for deuterides than for molecules composed of heavier atoms, while the values for hydrides are approximately double those for deuterides. The halomethanes show striking differences between different vibrational modes: for CH_3Cl the value for the C-Cl stretching mode, ν_3, is an order of magnitude lower than for all other modes: for $CHCl_3$ the effect is reversed, and the C-H stretching mode, ν_4, shows the largest value. This gives expression to the physical expectation that the efficiency of coupling between vibration and translation will increase with increasing amplitude of the vibration.

The magnitude of the whole matrix element $[V^{(i+1)\text{-}i}]^2$ corresponding to a one-quantum jump varies directly with α^2 and $(\overline{A^2})$, and inversely with the vibration frequency. It will be seen from Table 3.2 that the values of α^* do not vary greatly between different molecules, so that the determining factor is $(\overline{A^2})/\nu$. Thus comparing hydrides with deuterides, the lower vibration frequencies of the deuterides, which promote more efficient energy transfer, are partly offset by their lower amplitude coefficients, which have the opposite effect. A similar situation arises for the halomethanes where the higher frequencies of

TABLE 3.1.

Vibrational amplitude coefficients

Gas	Mode	Frequency (cm^{-1})	$(\overline{A^2})$ (a.m.u.$^{-1}$)
H_2	ν	4160	1·0
D_2	ν	2989	0·5
O_2	ν	1554	0·0625
Cl_2	ν	557	0·0282
NO	ν	1876	0·0672
HCl	ν	2885	0·9720
DCl	ν	2040	0·4735
H_2O (ref. 29)	ν_1	3657	0·9539
	ν_2	1596	0·9527
	ν_3	3756	0·9241
D_2O (ref. 29)	ν_1	2671	0·4611
	ν_2	1178	0·4373
	ν_3	2788	0·4374
SO_2 (ref. 87)	ν_1	525	0·0533
	ν_2	1150	0·0544
	ν_3	1360	0·0491
CH_4 (ref. 88)	ν_1	2915	0·9921
	ν_2	1534	0·9921
	ν_3	3019	0·9923
	ν_4	1306	0·8368
CH_3Cl (ref. 88)	ν_1	2966	0·9564
	ν_2	1355	0·8740
	ν_3	732	0·0939
	ν_4	3100	0·8949
	ν_5	1455	0·9364
	ν_6	1015	0·8485
$CHCl_3$ (ref. 5)	ν_1	3033	
	ν_2	667	0·0702
	ν_3	364	0·0366
	ν_4	1215	0·9709
	ν_5	760	0·0116
	ν_6	260	0·0319
CCl_4 (ref. 5)	ν_1	458	0·0043
	ν_2	218	0·0282
	ν_3	776	0·0261
	ν_4	314	0·0282

The vibrational modes are numbered according to the usual spectroscopic convention (see ref. 89).

modes involving H atoms are to some extent offset by their higher amplitude coefficients. The vibrational factors for double quantum jumps (3.6) are smaller than for single jumps and the same applies to higher multiple jumps, so that the transition from v to $(v-1)$ will always be the most probable. For a single quantum jump (3.5) the vibration factor increases proportionately to $(i+1)$, so

TABLE 3.2.

SSH calculations for diatomic molecules

Gas	T(K)	α^* (Å^{-1}) (method B)	Z_{10} (calc.) (c)
N_2	476	4·999	3·205 (9)
$\nu = 2331$ cm^{-1}	1000	4·961	1·942 (7)
(a)	2000	4·939	2·717 (5)
	5000	4·929	2·577 (3)
CO	286	5·114	1·47 (10)
$\nu = 2143$ cm^{-1}	500	5·070	3·86 (8)
(a)	1000	5·029	4·59 (6)
	1500	5·012	4·17 (5)
	2200	5·000	5·16 (4)
	3000	4·993	1·09 (4)
	5000	4·987	1·10 (3)
O_2	303	5·377	7·30 (7)
$\nu = 1554$ cm^{-1}	600	5·314	1·73 (6)
(a)	1372	5·263	2·51 (4)
	1953	5·247	5·12 (3)
	3000	5·233	8·40 (2)
	4000	5·222	1·30 (2)
Cl_2	200	5·057	1·44 (5)
$\nu = 577$ cm^{-1}	290	4·988	4·13 (4)
(b)	350	4·954	2·09 (4)
	435	4·917	9·35 (3)
	528	4·884	4·61 (3)
	1000	4·787	4·70 (2)

(a) Values for N_2, CO, and O_2 calculated by Stretton[23] using eqn (3.2).
(b) Values for Cl_2 and Br_2 calculated by Dickens and Ripamonti[113] using eqn (3.3).
(c) Values of Z_{10} expressed as $A\,(B) = A \times 10^B$.
Intermolecular force constants used in these calculations are given in Table 1.1.

that higher vibrational levels are more easily deactivated to translation: this effect is seldom observed experimentally, since upper levels usually deactivate very rapidly by near-resonant V-V transfer to the (v=1) level. (But see § 4.5).

(d) *The translation factor* gives expression to the change in kinetic energy of translation involved in the inelastic collision. The parameters involved are the reduced mass of the collision, μ, and the temperature T, which together determine the velocity of approach; ΔE, the amount of vibrational energy transferred in the collision, and the intermolecular repulsion parameter, α. It will be noted that all the terms involve ΔE^2 with the exception of the second exponential term, exp $(-\Delta E/2kT)$. This alone will change sign for the reverse process, $P_{l-k(\mathrm{b})}^{j-i(\mathrm{a})}$, and the two probabilities will differ by a factor exp $(-\Delta E/kT)$ in accordance with the principle of detailed balancing (§ 1.4.2).

The exponential term of the translation factor controls the temperature-dependence of P, since the whole the pre-exponential part of eqn (3.3) is only weakly temperature-dependent. The dominant term is usually the negative first exponential term, exp (-3ζ). This predicts that log P will vary with $T^{\frac{1}{3}}$ as in the Landau-Teller expression (1.4). But for cases where the attractive potential (represented by $-\epsilon$) is large the positive third exponential term exp $(+\epsilon/kT)$ can become dominant, giving rise to a higher transition probability which will vary *inversely* with temperature. The SSH theory thus predicts qualitatively the minimum in the Landau-Teller plot (log P versus $T^{\frac{1}{3}}$) which is observed for strongly polar molecules (cf. § 1.6), and the curvature observed for all other molecules at lower temperatures. Quantitative prediction under these conditions is unsuccessful due to the failure of eqn (3.1) to give adequate expression to the effect of the attractive potential, and the SSH theory consistently underestimates P for strongly polar molecules and at lower temperatures.

3.2. Modified theories

3.2.1. The defects of SSH theory

Two fundamental defects of the SSH theory have been pointed out above:

(a) Inadequate expression is given to the *attractive* part of the intermolecular potential. This becomes important for all molecules at low temperatures, and for strongly polar molecules at all temperatures.

(b) The relative orientation of the molecules is regarded as constant during the collision, and the possibility of simultaneous rotational transitions is disregarded. These assumptions are valid for molecules with large moments of inertia, but not for hydrides and deuterides.

The following two sections will discuss modifications which take these factors into account.

3.2.2. The attractive potential.

Better account may be taken of the attractive forces for non-polar (approximately isotropic) molecules by using the Morse double exponential potential:

$$V(r) = \epsilon \exp[-\alpha'(r-r_{\mathrm{m}})] - 2\epsilon \exp[-\tfrac{1}{2}\alpha'(r-r_{\mathrm{m}})]; \qquad (3.7)$$

this allows an analytical solution, and leads to an alteration in the exponential factor in eqn (3.3) which may be written[69]

$$P(T) = A(T) \exp\left[-3\zeta + \frac{4}{\pi}\sqrt{\frac{\epsilon\zeta}{kT}} + \frac{16\epsilon}{3\pi^2 kT} + \frac{\Delta E}{2kT}\right], \qquad (3.8)$$

where $A(T)$ represents the whole, weakly temperature-dependent pre-exponential factor in eqn (3.3). This assumes a spherically symmetrical potential and gives a good prediction for the temperature-dependence of P for simple non-polar molecules at lower temperatures. Eqn (3.8) is due to Shin:[69] a similar calculation for N_2, O_2, and CO has been published by Calvert and Amme.[90] The method can only be used when sufficient transport data are available to obtain the force constants for eqn (3.7).

For polar molecules, which are strongly anisotropic, the angle dependence of the intermolecular potential must be taken into account, and also the predominance of preferred collisional orientations at lower temperatures. This requires modification of *both* the pre-exponential *and* the exponential terms in eqn (3.3). Shin[69] has derived an expression

$$P(T) = \frac{A}{16(a^2-b^2)} \zeta^{\frac{3}{2}} \exp\left[-3\zeta + \frac{4}{\pi}\sqrt{\frac{\epsilon\zeta}{kT}} + \frac{16\epsilon}{3\pi^2 kT} + \frac{\Delta E}{kT} + 2a\right] \quad (3.9)$$

where

$$A = \sqrt{\frac{4\pi}{3}}\left(\frac{4\pi m\alpha\Delta E}{\hbar^2}\right)^2 \mathbf{X}^2 \quad (\mathbf{X}\text{ is the matrix element})$$

$$a = \frac{\mu^2}{\epsilon\sigma^3}\left(\frac{\epsilon^{\frac{3}{4}}\zeta^{\frac{1}{4}}}{(kT)^{\frac{3}{4}}}\right), \quad b = \left(\frac{\mu Q}{\epsilon\sigma^4}\right)\left(\frac{\epsilon^{\frac{2}{3}}\zeta^{\frac{1}{3}}}{(kT)^{\frac{2}{3}}}\right),$$

taking into account both dipole-dipole and dipole-quadrupole forces, (μQ). This expression contains an additional positive exponential term $2a$, which also varies inversely with temperature and will be important for molecules with high dipole moment and small molecular diameter, such as NH_3. It predicts with fair accuracy the inversion of the temperature-dependence of $P(T)$ observed at lower temperatures for molecules such as SO_2 (see Fig. 3.9).

3.2.3. Simultaneous rotational transitions: V-R and V-(T, R) transfer

Interaction between molecular vibration and rotation was proposed by Cottrell *et al.*[131] in order to account for the observation that many deuterides show lower values of P_{10} than the corresponding hydrides, in spite of their substantially lower vibrational frequencies (which in simple SSH theory would correspond to faster relaxation). Cottrell suggested that, since the actual rotational velocity of the hydrogen atoms in most hydrides is larger than the translational velocity of the molecule, vibrational energy transfer into rotation will be more probable than transfer into translation. He produced a quantitative theory based on the model of interaction of a classical rotator and a quantized oscillator, which was further developed and applied to a wide range of molecules by Bradley Moore.[132] It gives an expression for P_{10} which formally resembles the SSH expression (3.3), but with the important difference that, since the *angular velocity* of rotation rather than the translational velocity of the molecules plays the important role, ωd is substituted for the translational velocity and I/d^2 for the reduced mass (ω is the angular velocity, d the bond length of the hydride molecules, and I the moment of inertia). The dominating negative exponential term ζ of the SSH eqn (3.3), which may be written in the abbreviated form of eqn (3.8) as

$$P(T) = A(T)\exp(-3\zeta) = A(T)\exp\left[-3\left(\frac{\Delta E^2\mu\pi^2}{2\alpha^{*2}\hbar^2 kT}\right)^{\frac{1}{3}}\right] \quad (3.10)$$

is thus replaced by a new term giving (eqn 3 of ref. 132)

$$P(T) = A(T)'' \exp\left[-3\left(\frac{2\pi^2 I \Delta E^2}{d^2 \alpha^2 kT}\right)^{\frac{1}{3}}\right] \quad (3.11)$$

The gross effect of this change on the deuteride/hydride ratio may be seen from a simple example—CD_4/CH_4. The ratio of moments of inertia is $I_{CD_4}/I_{CH_4} = 6{\cdot}38/3{\cdot}19$, which is much larger than the ratio of molecular weights, $M_{CD_4}/M_{CH_4} = 20/16$, and this counteracts the favourable frequency ratio, $\nu_{CD_4}/\nu_{CH_4} = 2090/2886$, so that CH_4 relaxes faster (see Table 3.4). This theory, which treats vibrational relaxation as *entirely* due to V-R transfer, gives good prediction of the deuteride/hydride *ratio*, but predicts absolute rates which are generally too low (see § 3.4.1).

More recently Shin[133] has developed a classical collision theory for vibrational energy transfer in hydrogen halides. This solves the classical equations for the collision model of a 'rotation-averaged oscillator and a rigid rotator, in which two hydrogen atoms interact strongly at close-in collisions'. It takes account of the effects of *both* translational *and* rotational motion (i.e. of V-T and V-R transfer occurring simultaneously), and gives a good fit with experimental measurements for HCl, DCl, HBr, and HI over the temperature range 800-2000 K, where a 'normal' Landau-Teller temperature dependence is predicted and found (see Fig. 3.7).

Both the 'pure' semiclassical V-R theory and Shin's classical V-(T, R) theory treat the *rotational* motion of the molecule as classical. This seems to be justified for molecules, such as the hydrogen halides, where the quantum vibrational level spacings are very much larger than the quantum rotational level spacings, but it breaks down for polyatomic molecules having much lower vibrational frequencies. If both energy modes are regarded as quantized, two competing factors have to be taken into account. A ΔJ transition, simultaneous with, and in the opposite direction to a Δv transition, will exert a favourable effect on the transition probability by decreasing ΔE_{int}, the amount of internal energy transferred to translation, since

$$\Delta E_{int} = \Delta E_{vib} - \Delta E_{rot}.$$

This will tend to favour a large value of ΔJ and of ΔE_{rot}. This effect is balanced by the necessity for conservation of total angular momentum in collision, which favours a small value of ΔJ. Various theoretical calculations have been made, and all predict that a $\Delta v = \pm 1$ transfer will normally be accompanied by ΔJ transition of size depending on the mass and moment of inertia of the molecule. For a heavy molecule, such as I_2, the momentum requirements are more easily met, and ΔJ values up to ± 14 (with an optimum at 6 to 8) are predicted theoretically and confirmed experimentally.[195] For lighter molecules such as O_2 and N_2, $\Delta J = \pm 2$ is the most probable transition, being 50 per cent more probable than $\Delta J = 0$, and three orders of magnitude more probable than $\Delta J = \pm 4$.[188] (The even numbers are required to maintain the symmetry of the total wavefunction for a homonuclear diatomic molecule.) This is confirmed by experimental work on Li_2.[190]

A great many more rigorous and sophisticated theoretical treatments of vibrational energy transfer have been published, especially for simple systems such as He + H_2. They have not been widely used in the calculation of actual transition probabilities for comparison with experiment, and will only find mention in this book with reference to specific experimental situations. Extensive theoretical reviews have been given by Takayanagi[84] in 1965, by Rapp and Kassal[85] in 1969, and by Secrest[86] in 1973. The recently published book *Molecular collision theory* by M. S. Child (Academic Press, 1974) surveys the whole field. An entirely novel approach, using information theory, has recently appeared. This will be briefly discussed in § 3.6, with explicit reference to empirical correlations which appear earlier.

3.3. Relaxation in pure diatomic gases

3.3.1. Non-polar molecules: $N_2, O_2, Cl_2, CO, H_2, D_2$

Vibration-translation transfer in the first four gases has been very fully investigated both experimentally and theoretically. All show comparatively weak intermolecular attractive forces (see Table 1.1), and possess comparatively large moments of inertia, so that rotational effects will be unimportant, and the simple SSH theory should apply. They are all stable over a large temperature range, so that shock-tube measurements can be made at high temperatures. Measurements by acoustic techniques at lower temperatures are less accurate, owing to the small contribution of vibrational energy to the specific heat, but laser fluorescence measurements are now becoming possible. For the theoretical calculations reliable intermolecular force constants, based on viscosity measurements are available. Table 3.2 shows values of the collision number, Z_{10} ($=1/P_{10}$), calculated by the SSH method. These are shown on Landau-Teller plots, in comparison with representative experimental data, in Figs. 3.2, 3.3, 3.4, 3.5. All four show the predicted linear relation between log Z_{10} and $T^{-\frac{1}{3}}$ at higher temperatures, with definite curvature setting in at lower temperatures where the role of attractive forces in increasing the transition probability becomes apparent. The SSH calculations predict both the absolute values and the temperature-dependence remarkably well in the linear region, except for O_2 where Stretton's calculated values of Z_{10} are consistently lower than the experimental values by a factor of 0·5. In view of the uncertainty involved in the use of the available force constants, *both* in calculating the theoretical transition probabilities *and* in deriving experimental values of Z_{10} from observed relaxation times, better agreement could not have been expected.

The experimental data at lower temperatures are more scattered, but there is clear evidence of the expected curvature due to the increasing effect of intermolecular attraction. In general the SSH theory underestimates the deviation from linearity (i.e. predicts too low a transition probability) in this region, while the calculations of Shin[91] give better agreement. Shin predicts greatly increased curvature at still lower temperatures: at 100 K the ratio $Z_{10}\,(T)_{\text{Landau-Teller}}/Z_{10}\,(T)_{\text{Shin}}$ will be 255 for O_2, 221 for N_2, and 390 for CO: at 200 K its values

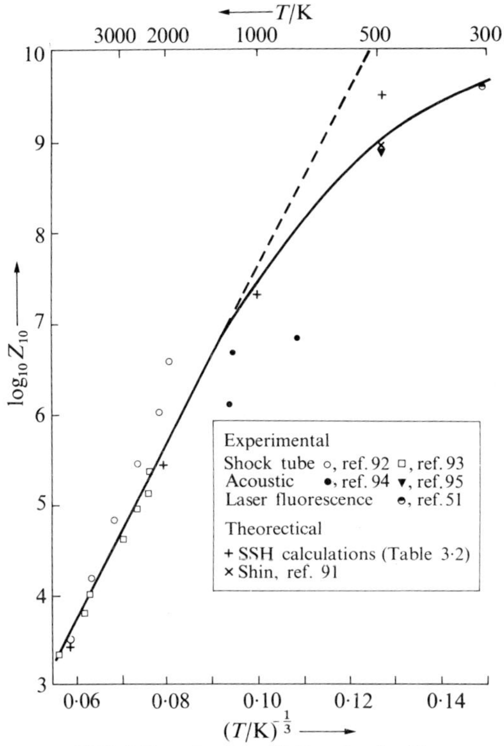

FIG. 3.2. Landau-Teller plot for N_2

are 33·0 for O_2, 31·6 for N_2, and 44·5 for CO. It is particularly important that these deviations should be taken into account in extrapolating transition probabilities to lower temperatures for astrophysical purposes. For Cl_2 (Fig. 3.5) the SSH calculations agree better with experiment at 303 K than Shin's calculation, so prediction to lower temperatures must be regarded as uncertain. It will be noted that the main factor influencing both the slope of the Landau-Teller plot and the absolute value of Z_{10} is ΔE, the size of the vibrational quantum interchanging with translational energy. Both slopes and absolute values in Figs 3.2, 3.3, 3.4, and 3.5 decrease in the order N_2 (2331 cm^{-1}) > CO (2143 cm^{-1}) > O_2 (1554 cm^{-1}) > Cl_2 (557 cm^{-1}), as predicted by eqn (3.2). The role of mass is less important, and the values of α^* shown in Table 3.2 do not vary much between the different molecules. SSH calculations for temperatures higher than those quoted are unreliable because of breakdown of perturbation theory, and consistently underestimate the collision number. For heavier molecules with very small vibrational quanta, multiple quantum transitions need to be taken into account. Kurzel and Steinfeld[251] have investigated the fluorescence from laser-excited $I_2\,(B^3\,\Pi^+_{ou})$ and found evidence for vibrational transitions ranging from $\Delta v = +5$ to $\Delta v = -8$ out of the $v' = 43$ level.

Vibrational relaxation in H_2 (ν = 4160 cm^{-1}) and D_2 (ν = 2989 cm^{-1})

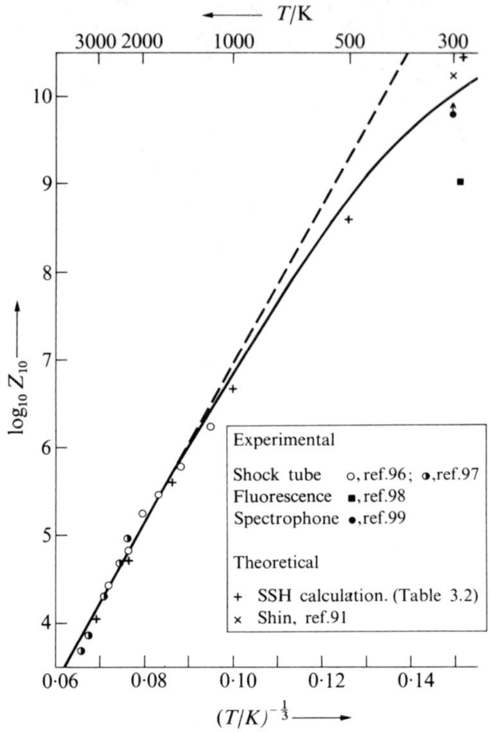

FIG. 3.3. Landau-Teller plot for CO

presented experimental difficulties because of very small vibrational contribution to the specific heat, except at high temperatures, which makes acoustic measurements impossible. Earlier shock-tube measurements were unreliable due to the low density and refractive index of the gas. There are now reliable shock-tube measurements, made in the range 1100 to 3000 K using laser-Schlieren techniques,[114,115] and accurate laser-Raman measurements[116,118] over the range 40 to 500 K for both H_2 and D_2. Landau-Teller plots of log Z_{10} calculated from these measurements are shown in Fig. 3.6. Both show a linear relation between log Z_{10} and $T^{-\frac{1}{3}}$ over the range 300 to 3000 K, with marked curvature corresponding to increased transfer probability setting in at lower temperatures. This behaviour is analogous to that shown in Figs 3.2, 3.3, and 3.4, but owing to the low mass and high translation velocity of both gases, SSH-type theory is not applicable quantitatively. Predictions made by a numerical three-dimensional distorted-wave calculation due to Calvert,[117] which use a Morse potential for interaction between all four atoms and ignore any contribution from V-R transfer are shown on Fig. 3.6 and give good approximations for both molecules at temperatures above 300 K.

The fact that Z_{10} values for D_2 are consistently higher than those for H_2, in spite of the much lower vibrational frequency of D_2, shows that the high

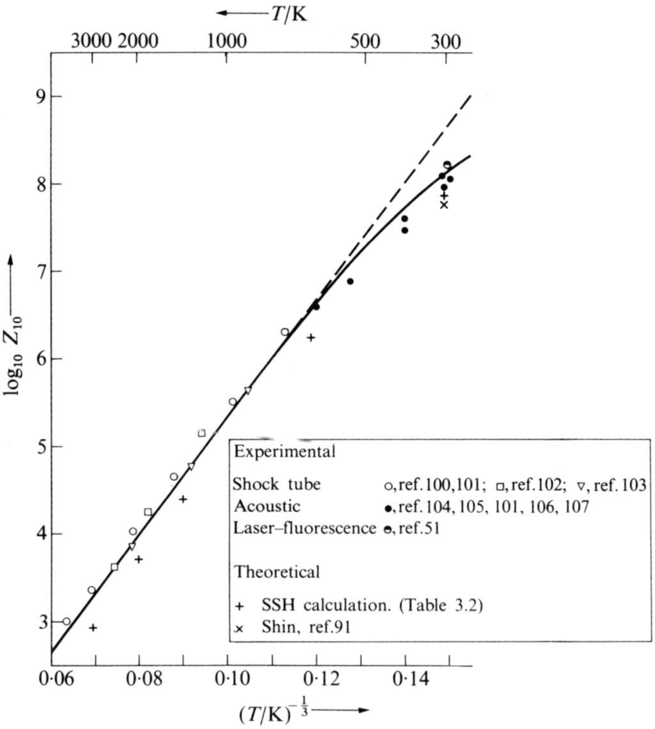

FIG. 3.4. Landau-Teller plot for O_2

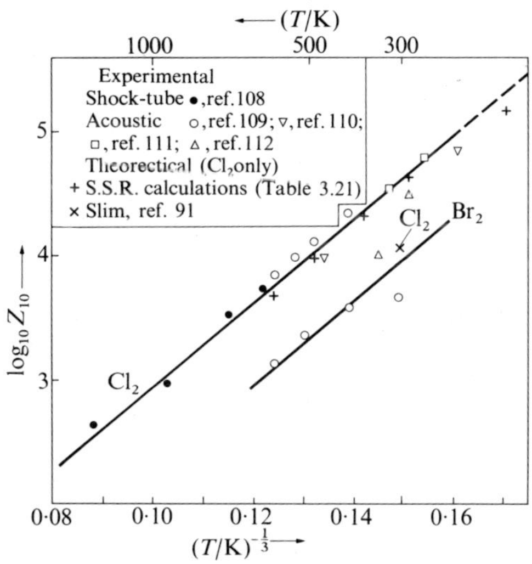

FIG. 3.5. Landau-Teller plots for Cl_2 and Br_2

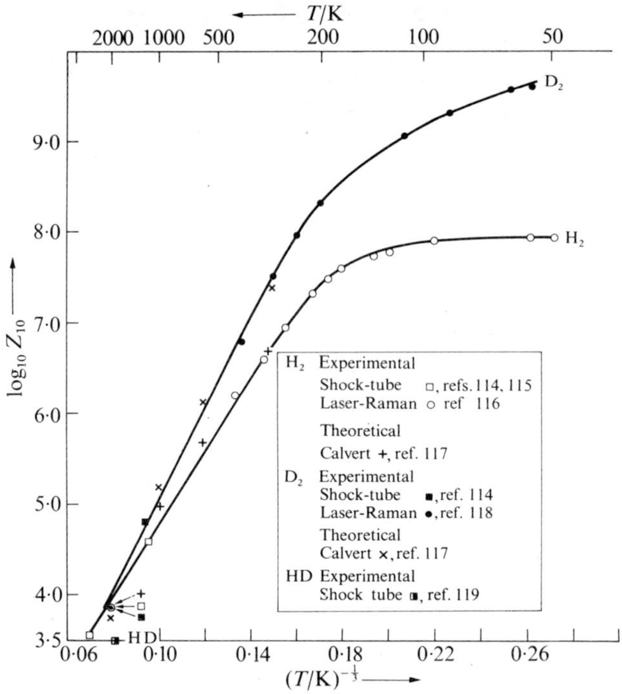

FIG. 3.6. Landau-Teller plots for H_2, D_2, and HD

translational velocity and vibrational amplitude of H_2 play a determining role. This also accounts for the V-T transfer probability for H_2 at 1000 K being two orders of magnitude larger than for N_2 or CO. A simple physical explanation may be given of the apparent constancy of Z_{10} for H_2 at temperatures below 50 K. The attractive potential energy constant for H_2 is $\epsilon = 75{\cdot}5$ cal mol^{-1} (Table 1.1) so that at temperatures $\leqslant$ 38 K, where $\epsilon \geqslant RT$, the impact energy for all collisions becomes effectively equal to ϵ and remains constant at this value, which will determine the transfer probability. There will be no temperature-dependent orientation effects in this range, as the rotational specific heat of H_2 is effectively zero. This is in contrast to the behaviour of the hydrogen halides discussed in the following section, where orientation effects produce an *inversion* of temperature-dependence at lower temperatures.

The experiments quoted above were all performed with n-H_2. Recent low temperature work on *o*-H_2 and *p*-H_2 mixtures of differing composition[120] has shown that, as temperature falls below 200 K, *o*-H_2 relaxes increasingly faster than *p*-H_2, while no difference is observed at higher temperatures. This suggests that V-R transfer must begin to play a significant role at low temperatures, and a theoretical treatment by Sorensen[121] on these lines gives good agreement with experiment (though it fails for D_2[120]). A single shock-tube result[119] for HD, which is plotted on Fig. 3.6, shows that at 2000 K the V-T transfer probability for HD is larger than for either H_2 or D_2. This would seem to indicate that rotational participation is also significant at high temperatures.

3.3.2. Polar molecules: H and D halides

The potential and molecular constants for hydrogen and deuterium halides are shown in Table 3.3. These molecules form an excellent test of the modifications of the SSH theory discussed in § 3.2. All are strongly polar, with values of the dipole moment μ ranging from 0·42 to 1·92 Debye. They show correspondingly large values of the attractive parameter ϵ, and HF and HCl both form specifically oriented dimers, with values of the heat of dimerization, $-\Delta U_{dim}$, of 6000 and 2100 cal mol^{-1} respectively. All have small moments of inertia I, those for the hydrides being smaller than for the deuterides. Vibrational relaxation times have been measured by shock-wave and laser fluorescence techniques over a large temperature range: the values of $\log Z_{10}$ are shown on Landau-Teller plots in Fig. 3.7.

TABLE 3.3.

Potential and molecular constants of hydrogen halides

Gas	Molecular weight	I (10^{-40}g cm^2)	ν_{10} (cm^{-1})	μ (10^{-18} esu)	σ (10^{-8} cm)	ϵ (cal mol^{-1})	$-\Delta U_{dim}$ (cal mol^{-1})
HF	20·0	1·34	3961	1·92	2·5*	795*	6000 (ref. 135)
DF	21·0	2·54	2906				
HCl	36·5	2·64	2886	1·08	3·3	715	2100 (ref. 136)
DCl	37·5	5·14	2090				
HBr	80·9	3·30	2560	0·8	3·4	828	
DBr	81·9	6·53	1840				
HI	127·9	4·27	2230	0·42	4·1	643	
DI	128·9	8·60	1599				

*Values from transport data are not available. $\sigma = 2·5$ Å is the estimated F-H ··· F- distance in the HF polymer,[9] and gives the best fit to relaxation data. $\epsilon = 795$ cal mol^{-1} was estimated from other physical data by Bott and Cohen.[122]

The qualitative picture is exactly as predicted. All show a 'normal' Landau-Teller temperature dependence at higher temperatures. As the temperature falls, Z_{10} reaches a maximum value and then decreases as oriented dipole-dipole forces become more effective in collisions. It is interesting that the ratio of $T_{max} \sim 1000$ K for HF to $T_{max} \sim 350$ K for HCl is almost exactly equal to the ratio of the heats of dimerization, 6000 cal mol^{-1} for HF to 2100 for HCl. Values of Z_{10} are consistently higher for the deuterides than for the corresponding hydrides in spite of lower vibrational frequencies, indicating that V-R transfer plays an important role. V-R transfer also accounts for the fact that HCl and HBr show almost identical values of Z_{10}, in spite of a mass ratio

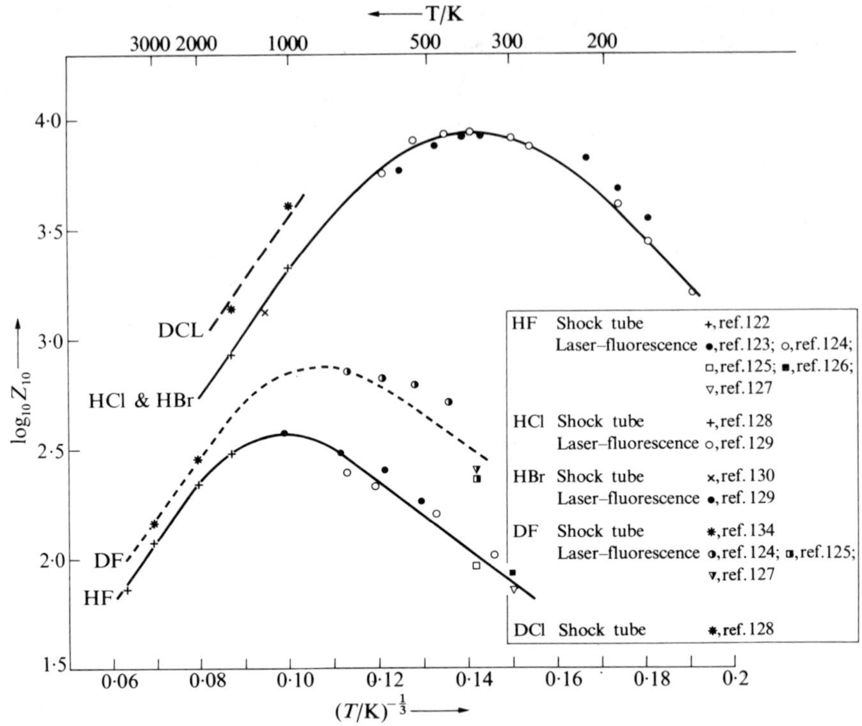

FIG. 3.7. Experimental Landau-Teller plots for HF, HCl, HBr, DF, and DCl

HBr/HCl > 2. The moment of inertia of HCl, $I = 2{\cdot}64 \times 10^{-40}$ g cm^2, is only slightly smaller than that of HBr ($I = 3{\cdot}30$) and, in the context of eqn (3.11), this just compensates for the small drop in vibrational frequency (2886 to 2650 cm^{-1}) between HCl and HBr. (Data are also available for HI and DI.[130,137] They differ only slightly from data for HCl and HBr, and have been omitted from Fig. 3.7 to avoid confusion.) HF, with a much higher vibrational frequency, $\nu = 4138$ cm^{-1}, has a smaller molecular weight, a substantially smaller moment of inertia, and much more powerful oriented dipole-dipole interaction, as shown by the large heat of dimerization. These factors combine to produce a vibrational relaxation rate which is several orders of magnitude faster than the rate for hydrogen, and much faster than for HCl and HBr.

Quantitative theoretical calculations, which may be found in the experimental references quoted, show that a fair approximation to the experimental data may be predicted by using the V-R theory of Shin[133] (§ 3.1.4), modified so as to include the effects of the strong dipole-dipole interaction. Bott and Cohen[134] obtain a fair approximation for HF and DF using the Shin theory with an attractive potential, $\epsilon = 3{\cdot}5$ kcal mol^{-1}, corresponding to half the energy of dimerization. Shin[138] has made detailed calculations for HF on the assumption that the intermolecular forces are a combination of a Morse potential with oriented dipole-dipole interaction, and obtains a good fit with experiment over the entire temperature range. Calculations for HCl and HBr are less satisfactory.[129] The 'pure' V-R theory of Bradley Moore predicts values of Z_{10} for HF and DF which are much higher than the experimental ones.[122]

3.3.3. *Vibronic transitions: NO*

NO is a *weakly* polar molecule; $\mu = 0{\cdot}153$ Debye, and the Lennard-Jones collision parameters, $\sigma = 3{\cdot}6$ Å, $\epsilon = 181$ cal mol^{-1}, and molecular weight 30 closely resemble those of O_2 and CO. The vibration frequency, 1878 cm^{-1}, lies between CO and O_2. But, in contrast, it forms a dimer by chemical bonding for which estimates of $-\Delta U_{\text{dim}}$ vary between 1600 and 3600 cal mol^{-1}, with an oriented structure[139,140]

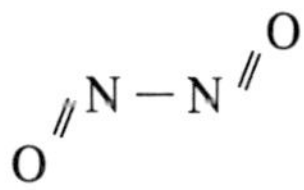

This resembles the HCl dimer ($-\Delta U_{\text{dim}} = 2100$) in binding energy. Experimental relaxation times have been measured at temperatures > 1500 K by shock tube,[141,142] and show a normal Landau-Teller temperature dependence: values of Z_{10} range from 1000 at 1500 K to 316 at 3000 K. Acoustic measurements between 498 and 423 K follow the same temperature dependence, Z_{10} rising to 5200 at 423 K.[143] A single acoustic measurement[144] at 296 K gives the lower value $Z_{10} = 2700$, so that Z_{10} would appear to rise to a maximum value around 6000 at a temperature around 350 K. Flash photolysis measurements in NO + N_2 mixtures confirm a negative temperature dependence at temperatures between 300 and 100 K.[145] The qualitative relaxation behaviour closely resembles that of HCl shown in Fig. 3.7. $T_{\max}$ is roughly the same, but the absolute values of Z_{10} are consistently smaller than for HCl by a factor of roughly 0·5. It seems that the reversal of temperature dependence is dominated by the dimerization process in much the same way as for HCl, but this does not explain why NO relaxation is so much faster than O_2 or CO at higher temperatures where orientation-dependent forces become less important. NO is *not* a hydride, so that V-R transfer cannot be invoked, and some other explanation must be sought.

NO is the only diatomic molecule possessing a doublet electronic ground state, $^2\Pi_{\frac{1}{2}}$ and $^2\Pi_{\frac{3}{2}}$. The corresponding energy levels differ by 0·085 R at 296K, and this gives rise to a collisional relaxation of the electronic, in addition to the vibrational and rotational degrees of freedom. Bauer and Sahm[144] made acoustic absorption measurements over a wide range of f/p and found:

$$
\begin{aligned}
Z_{10}\ &(\text{vibrational}) &= 2700\\
Z\ &(\text{electronic}) &= 70\\
Z\ &(\text{rotational}) &= 1.
\end{aligned}
$$

In addition the ground state NO $X^2\Pi$ molecule possesses resultant orbital angular momentum, and Nikitin[146] postulated a *vibronic* transition, in which vibrational energy transfer is accompanied by a change in the electronic state of the system, as responsible for its unexpectedly efficient vibrational relaxation. In a collision between two NO $X^2\Pi$ molecules there is a potential splitting, and resonance occurs between the vibrationally excited ground state, $^3\Sigma_g^-$, and the vibrationless $^1\Sigma_g^+$ state. Nikitin's theory predicts a positive temperature dependence, but Billingsley and Callear[145] suggest that the observed inversion of temperature-dependence is due to the resonance energy being dependent on orientation. The problem is complicated by the additional possibility of actual dimer formation. Detailed discussion will be found in ref. 145.

3.4. Relaxation in pure polyatomic gases

3.4.1. The limitations of present theory

The two major characteristics affecting the comparison between relaxation behaviour of polyatomic gases and of diatomics are:

(a) Polyatomic molecules possess several vibrational modes, each of which contributes to the molecular vibrational heat capacity, and can, in principle, relax independently.

(b) Polyatomic molecules are structurally flexible and change their effective shape both during normal vibration, and, more important, during collision.

Consideration (a) turns out to be relatively undisturbing. The vast majority of polyatomic molecules show a single vibrational relaxation process involving the whole vibrational heat content of the molecule. This is due to rapid intramolecular V-V transfer, which maintains continuous equilibrium of vibrational energy between the active fundamental modes. The whole vibrational heat content of the molecule then relaxes in a single V-T transfer *via* the lowest mode (which will usually have a higher V-T transfer rate than upper modes). This mechanism is characterized by a single overall relaxation time, β, which may be related to β_1, the actual relaxation time of the lowest mode, by the equation $\beta_1 = (C_1/C_s)\,\beta$, where C_s is the total vibrational heat content and C_1 the contribution due to the lowest mode.[153] (It has been pointed out by Shields[369] that this relation is valid only when the V-V rate is faster than the V-T rate, and when C_1 represents a major part of the vibrational specific heat. This is usually the case.) The intramolecular V-V transfer process will be considered in detail in Chapter 4. Here we shall consider mainly the characteristics of the V-T transfer from the lowest mode (cf. § 4.2.1).

The consequences of (b) are much more serious. For most polyatomic molecules even an approximately quantitative description of an elastic collision is at present impossible, while for inelastic collisions, first-order perturbation theory is inadequate. In consequence, the kind of theoretical treatment discussed in § 3.1 can only be applied qualitatively for most polyatomic molecules. The one gas for which quantitative calculations have been applied with some success is CH_4. This is an effectively spherical molecule, whose collision parameters (Table 1.1) are well known. The shape of the molecule will not be changed drastically by vibration, so that it may reasonably be regarded as a 'breathing sphere'. It shows a single relaxation time, characteristic of the lowest mode, $\nu_4 = 1306\ \text{cm}^{-1}$, which has been investigated experimentally over a wide temperature range. A Landau-Teller plot of experimental data is shown in Fig. 3.8, together with SSH calculations for V-T transfer for the ν_4 mode made

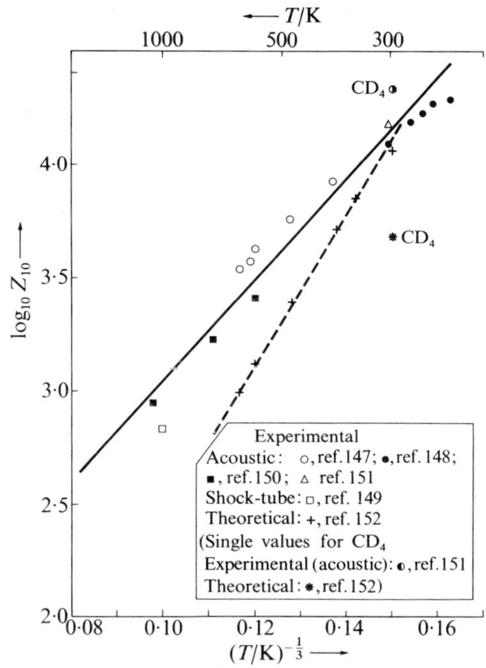

FIG. 3.8. Landau-Teller plot for CH_4

by Stretton.[152] Good agreement is found at 300 K, and both experimental and theoretical points show a linear plot, though the theoretical slope is significantly steeper. Calculations have also been made for other simple hydrocarbons and their derivatives by Stretton,[152] but they are less successful quantitatively. They give interesting qualitative information about the intramolecular V-V transfer processes, which will be discussed in Chapter 4. In contrast, attempts at quantitative theoretical treatment for a flexible linear molecule as simple as CO_2 are shown by recent shock-tube measurements[154] to be highly inaccurate, though they do predict correctly a single relaxation process for the total vibrational energy. Treatment of the intermolecular forces in a $CO_2 + CO_2$ collision is complicated by the fact that an instantaneous dipole moment is developed when the molecule is bent by oscillation or collision, and similar difficulties arise in other polyatomic molecules.

The effects of an oriented dipole-dipole interaction are similar for polyatomic and diatomic molecules, as may be seen by comparing the force parameters given for polar polyatomic gases in Table 1.2 with those for hydrogen halides in

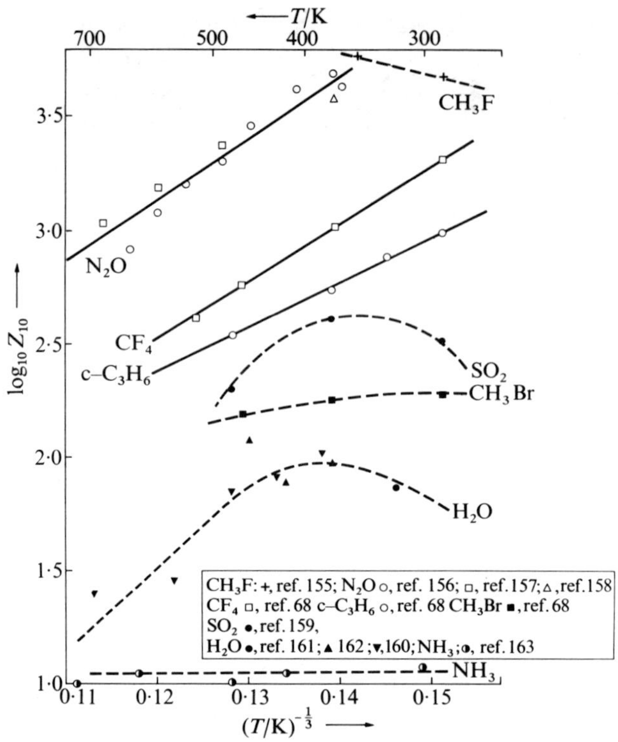

FIG. 3.9. Experimental Landau-Teller plots for polyatomic gases

Table 3.3. Fig. 3.9 shows the temperature-dependence of Z_{10} for some polyatomic gases which have been investigated over a sufficient temperature range. It will be seen that the non-polar gases, c–C_3H_6, CF_4, and the weakly polar gas N_2O ($\mu = 0{\cdot}167$ D) all show a normal positive Landau-Teller temperature-dependence, as was found in CH_4, with gradients increasing with increase in the absolute values of Z_{10}. (These gradients are not correctly predicted by SSH theory.) Of the strongly polar molecules, SO_2 which forms a dimer ($-\Delta U_{dim} = 4750$ cal mol^{-1}) shows a clear reversal of temperature-dependence, for which $T_{max} \sim$ 350 K, similar to that shown by HCl. H_2O, which also dimerizes ($-\Delta U_{dim} =$ 4980 cal mol^{-1}), shows a similar reversal: $T_{max} = 370$ K. (Recent shock-tube measurements[232] give a value $Z_{10} \sim 3$ at 2000 K, confirming the Landau-Teller temperature-dependence at higher temperatures.) The methyl halides, CH_3F, CH_3Cl, and CH_3Br have roughly equal dipole moments, $\mu \sim 1{\cdot}8$ D. CH_3F shows a definite negative dependence between 292 and 370 K, while CH_3Br shows a weak positive dependence between 291 and 456 K which may indicate a maximum at a slightly lower temperature. Measurements on pure CH_3Cl,[164] which are not plotted, show a normal Landau-Teller plot between 293 and 475 K, which lies close to that for c-C_3H_6. (Earlier measurements[68] showing a temperature inversion were made with impure gas.) NH_3, which forms a dimer

($-\Delta U_{dim}$ = 3680 cal mol^{-1}), shows a barely perceptible temperature dependence of its very low value of $Z_{10} \sim 10$ over the range 300 to 773 K. These results show rough qualitative agreement with the predictions of Shin.[69]

Deuteration of polyatomic hydrides produces the same effect on relaxation as was observed for the halogen hydrides (Fig. 3.7). The comparison of CH_4 (ν_{min} = 1306 cm^{-1}) with CD_4 (ν_{min} = 996 cm^{-1}) in Fig. 3.8 shows that, while the SSH theory for V-T transfer (which works fairly well for CH_4) predicts a *lower* value of Z_{10} for CD_4, the experimental value is substantially *higher* than for CH_4. Other hydrides behave in the same way, indicating that V-R transfer plays a role here also. Cottrell[131] has shown that his 'pure' V-R model predicts correctly the *ratio* of hydride/deuteride relaxation times, though prediction of the absolute values is not so successful. Table 3.4 shows the comparison.

TABLE 3.4

Ratios of vibrational relaxation times for deuterides and hydrides

Molecules (ν cm^{-1})	Theory β_D/β_H V-R	Theory β_D/β_H V-T	Expt β_D/β_H
CD_4 (996): CH_4 (1306)	1·7	0·58	1·7
SiD_4 (681): SiH_4 (914)	1·5	0·25	1·5
PD_3 (730): PH_3 (992)	1·5	0·13	1·5
AsD_3 (660): AsH_3 (906)	1·2	0·04	0·6

Data from ref. 131. Values of Z_{10} will be found in Tables 3.5 and 3.6.

Whether this simple semiclassical model, developed for diatomic hydrides, of a rapidly rotating hydrogen atom making an impact on the other molecule which is responsible for energy transfer, can be applied to a polyatomic hydride such as CH_4 is open to some doubt. While it is reasonable to regard the highly anisotropic HF molecule as a rapidly rotating dumb-bell, the potential energy surface surrounding a CH_4 or SiH_4 molecule will resemble the classical 'rough sphere', with the hydrogen atoms represented by only small pimples on the circumference. The fact that an SSH V-T calculation for CH_4 (Fig. 3.8) consistently predicts Z_{10} values rather *lower* than the experimental ones for both CH_4 and CD_4 does not accord with the view that a straight V-R transition is the more efficient and dominant mechanism. An alternative concept of a partial role played by V-R transfer is suggested in the following section.

3.4.2. An empirical correlation—the Lambert-Salter plot

An empirical correlation which fits most of the presently available experimental V-T relaxation rates for polyatomic gases at 300 K is the Lambert-Salter plot.[165,166] This is a simple plot of $\log_{10} Z_{10}$ against ν_{min}, the frequency of the lowest vibrational mode. Molecules fall into two classes, differentiated by the presence or absence of hydrogen atoms. The plot for hydrogen-containing

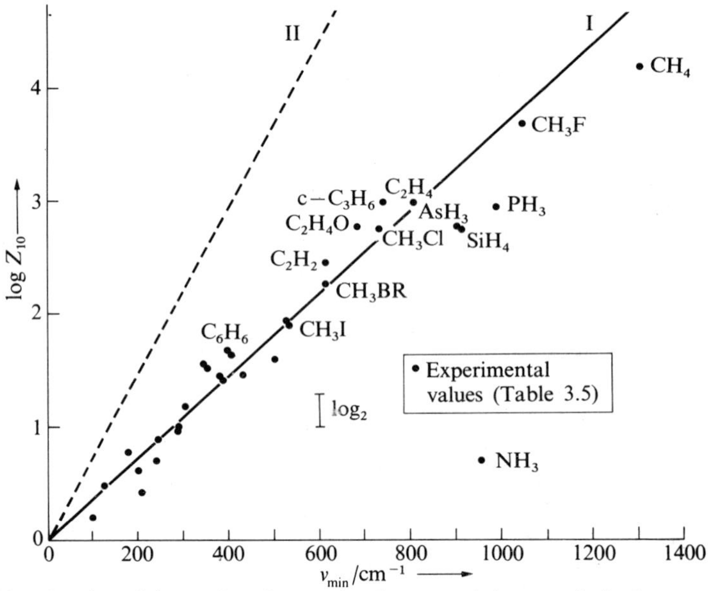

FIG. 3.10. Lambert-Salter plot for molecules containing ⩾ 2 hydrogen atoms.[166]

molecules is shown in Fig. 3.10; the data are listed (in sequence of their Z_{10} values) in Table 3.5. The plot for deuterated molecules is shown in comparison with the hydrogen-containing plot in Fig. 3.11: data in Table 3.6. The plot for all other molecules is shown in Fig. 3.12; data in Tables 3.7 and 3.8. It will be seen that, with a few exceptions, all molecules in each class show values of Z_{10} which fit a linear relation, $\log_{10} Z_{10} = X\, \nu_{min}$ within a factor of roughly 2.

TABLE 3.5.

Vibrational relaxation data at 300 K for polyatomic molecules containing ⩾ 2 hydrogen atoms (ref. 166)

Molecule	ν_{min} (cm^{-1})	β (ns)	Z_{10}	$\log_{10} Z_{10}$	Reference
CH_4	1306	2030	15160	4·18	151
CH_3F	1048	1900	4800	3·68	68
c-C_3H_6	740	311	990	2·99	68
C_2H_4	810	391	970	2·99	68
PH_3	991	196	890	2·95	167
C_2H_4O	685	360	603	2·78	168
AsH_3	905	310	605	2·78	131
CH_3Cl	732	204	575	2·76	169
SiH_4	914	115	570	2·76	151
C_2H_2	612	72·8	280	2·45	165
CH_3Br	611	34·3	190	2·28	68
CH_2F_2	529	30·0	83	1·92	169
CH_3I	533	37·0	78	1·89	170

Molecule	ν_{min} (cm^{-1})	β (ns)	Z_{10}	$\log_{10} Z_{10}$	Reference
CH_2CHCl	395	23·0	48	1·68	171
C_6H_6	404	29·0*	43	1·63	172
CH_2CHF	500	25·0	41	1·61	171
CH_2CHBr	345	24·0	36	1·56	171
CH_2:C:CH_2	354	5·2	33	1·52	33
C_2H_5OH	431	3·5	29	1·46	173
$CH_3COOC_2H_5$	376	8·5	29	1·46	173
CH_2ClF	385	10·0	26	1·42	169
CH_3COOCH_3	303	4·9	15	1·18	173
C_2H_6	290	15·0*	10	1·0	174
CH_2Cl_2	283	1·95*	9·6	0·98	175
C_3H_7OH	243	4·0	8	0·90	173
CH_3CH:CH_2	177	15·0	6	0·78	174
CH_3CHCl_2	240	5·35	5	0·70	176
NH_3	950	0·74	5	0·70	177
$C(CH_3)_4$	200	4·4	4·2	0·62	178
CH_2ClCH_2Cl	125	4·25	3	0·48	176
c-C_5H_{10}	207	1·9	2·6	0·42	178
n-C_4H_{10}	102	1·3	1·6	0·20	178
(n-C_5H_{12}	88	1·1	0·88	~0	178)
(n-C_6H_{14}	61	1·2	0·66	~0	178)

*values of β_1 for molecules showing multiple relaxation

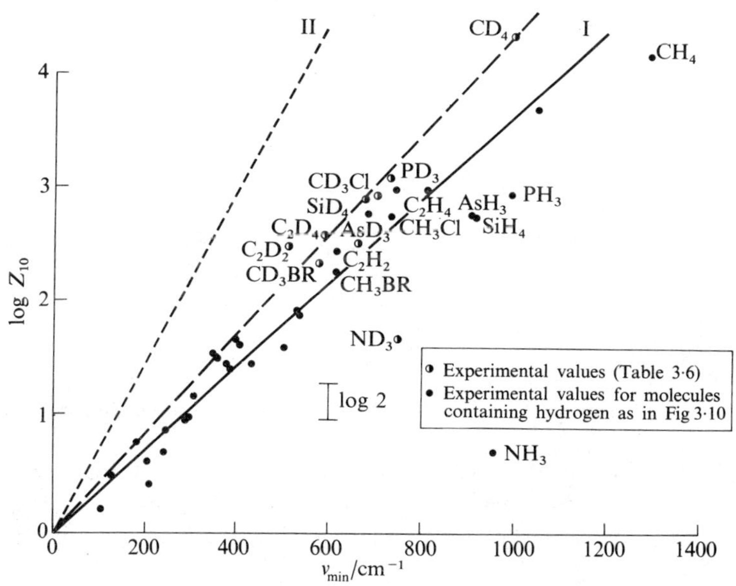

FIG. 3.11. Lambert-Salter plot for deuterated molecules[166]

TABLE 3.6.

Vibrational relaxation data at 300 K for deuterated polyatomic molecules (ref. 166)

Molecule	ν_{min} (cm^{-1})	β (ns)	Z_{10}	$\log_{10} Z_{10}$	Reference
CD_4	996	3900	21730	4·34	151
PD_3	728	343	1230	3·09	167
CD_3Cl	699	590	880	2·94	87
SiD_4	675	115	810	2·91	151
C_2D_4	584	140	390	2·59	179
AsD_3	660	220	337	2·53	131
C_2D_2	511	71	311	2·49	152
CD_3Br	577	142	221	2·34	165
ND_3	748	13	49	1·69	167, 177

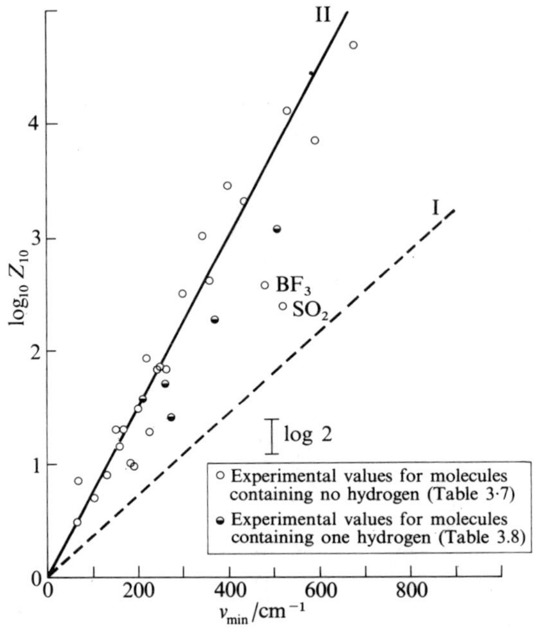

FIG. 3.12. Lambert-Salter plot for molecules containing ⩽ 1 hydrogen atom[166]

The inference follows that the effective probability of V-T transfer between v=0 and v=1 levels for each of these large classes depends, to a fair approximation, *only* on ΔE $h\nu_{min}$ the amount of vibrational energy transferred to translation, so that

$$P_{10} = 1/Z_{10} = \exp(-C\Delta E). \tag{3.12}$$

TABLE 3.7.

Vibrational relaxation data at 300 K for polyatomic molecules containing no hydrogen atoms (ref. 166)

Molecule	ν_{min} (cm^{-1})	β (ns)	Z_{10}	$\log_{10} Z_{10}$	Reference
CO_2	673	6370	53000	4·70	180
COS	527	1670	13000	4·12	147
N_2O	589	990	7320	3·86	181
CS_2	397	727	2950	3·47	168
CF_4	437	853	2070	3·32	68
SF_6	344	860	1005	3·02	182
CF_3Cl	356	239	427	2·63	169
BF_3	480	90	380	2·58	35
CF_3Br	297	196	316	2·50	169
SO_2	519	50*	250	2·40	183
CCl_4	218	19	86	1·93	184
$CFCl_3$	248	40	72	1·86	169
CF_2Cl_2	261	78	68	1·83	169
C_2Cl_4	237	57	68	1·83	185
CF_2BrCl	200	50	30	1·48	169
$SiCl_4$	150	5·9	20	1·30	184
C_2N_2	226	4·15*	18	1·28	183
CF_2Br_2	165	29	14	1·15	169
CF_3CN	187	3·18*	10	1·0	183
C_2F_4	190	17	9·6	0·98	171
$GeCl_4$	134	3·4	8	0·90	184
C_2F_6	68	28	7	0·85	165
$SnCl_4$	104	2·5	5	0·70	184
C_3O_2	63	1·0*	3	0·48	33

*values of β_1 for molecules showing multiple relaxation

TABLE 3.8.

Vibrational relaxation data at 300 K for Polyatomic molecules containing one hydrogen atom (ref. 166)

Molecule	ν_{min} (cm^{-1})	β (ns)	Z_{10}	$\log_{10} Z_{10}$	Reference
CHF_3	507	479	1230	3·09	169
$CHClF_2$	369	95	191	2·28	169
$CHCl_3$	260	13·5	52	1·72	175
$CHBrCl_2$	215	9·8	37	1·57	186
$CHCl_2F$	276	25	26	1·42	169
$CHBr_2Cl$	168	6·3	20	1·30	186

The values of the constant C when ΔE is expressed in cm^{-1} are

$C_I = 8{\cdot}3 \times 10^{-3}$ cm for the hydrogen-containing plot

$C_{II} = 17 \times 10^{-3}$ cm for the non-hydrogen-containing plot.

It is interesting to note that the plot includes flexible hydrocarbon molecules, such as n-C_5H_{12}, which possess very low torsional frequencies ($\nu_{min} < 100$ cm^{-1}) and that for these V-T transfer, like R-T transfer, occurs at practically every collision.

The striking exceptions, SO_2 and NH_3, which show *lower* values of Z_{10}, are both strongly polar molecules where the influence of dipole-dipole attraction has been shown in § 3.3.1 to dominate the relaxation process. H_2O[161] and H_2S,[191,192] which have higher frequencies and do not appear on the plot, fall into the same class. Both are strongly polar and show reversal of temperature-dependence (for H_2O see Fig. 3.9), and both show *low* values of Z_{10} at 300 K: H_2O ($\nu_{min} = 1595$ cm^{-1}) has $Z_{10} \sim 73$ and H_2S ($\nu_{min} = 1290$ cm^{-1}) has $Z_{10} \sim 290$. The other exception, BF_3, remains anomalous: the value is based on a single measurement. Diatomic molecules do *not* fit the correlation: Z_{10} values for Br_2 and Cl_2 lie higher than plot II, while the values for other diatomic molecules, which have much higher frequencies, lie below the plot. It is interesting that the strongly polar molecules CH_3Cl, CH_3Br, and CH_3I all conform closely to plot I. Table 1.2 shows that the *dimerization* energy for CH_3Cl is much lower than for SO_2, NH_3, or H_2O, and it would appear that this, rather than just the dipole moment, is the factor which is important in producing significant deviation from the behaviour of the other polyatomic molecules. The inference can be drawn that those polyatomic molecules whose relaxation is due to a predominantly *repulsive* interaction, and follows a normal Landau-Teller temperature-dependence, conform well to the Lambert-Salter plot.

Neither of the salient features of the plot, the dependence of Z_{10} solely on ΔE, together with the highly critical dependence on the presence of hydrogen atoms, can be given a fully satisfactory theoretical explanation at present. All theories predict an exponential dependence on ΔE, but other factors, mass etc., also play an important role and it is difficult to see why their influence should be effectively absent for such a wide variety of molecules. The suggestion made earlier,[165] that the probability of V-T transfer depends on the *kinetic energy* of approach in collision, predicts a different and much larger positive temperature-dependence than is observed, and is contrary to the physical principles discussed in Chapter 1. An entirely new theoretical approach, based on information theory, has recently been developed by Levine and Bernstein, and predicts a simple exponential dependence on ΔE for both vibrational and rotational relaxation. This will be discussed further in § 3.6. The special role of hydrogen atoms, due to their low mass leading to high vibrational amplitude, has also been pointed out in § 3.1.3 (c). Table 3.1 shows that the vibrational amplitude coefficients for CH_4 are larger than those for CCl_4 by a factor of approximately 30. But the vibrational factors appear in the pre-exponential part of eqns (3.2) and (3.3), and nothing suggests that the exponential factor C in eqn (3.12) should be halved by the presence of hydrogen atoms, leading to a decrease of several orders of magnitude in Z_{10} in the higher frequency range. The fact that polyatomic molecules containing only one hydrogen atom fall close to the non-hydrogen plot II in Fig. 3.12 suggests that some specific factor important in collisions between molecules, whose intermolecular potential depends on the interaction of peripheral shells made up predominantly of very light atoms, may be responsible. The plot for deuterated molecules in Fig. 3.11 falls a *little* above the hydrogen-containing, as might be expected in terms of this explanation.

The proponents of 'pure' V-R theory[131,132] suggested that the difference between plots I and II is due to the fact that *all* hydrogen-containing molecules relax exclusively by R-V transfer. This does not accord with the experimental data, since plot I clearly includes hydrogen-containing molecules which, because of their structure, possess no low moment of inertia: c-C_3H_6, C_2H_4O, C_2H_2, C_6H_6, c-C_5H_{10}. There are, however, *small* systematic deviations from plot I, which may be explained on the basis of V-(T, R) transfer, as discussed in the following section.

3.4.3. *The role of V-(T,R) transfer*

It was stated in § 3.2.3 that vibrational transitions $\Delta v = \pm 1$ in molecules of moderate molecular weight are usually accompanied by simultaneous rotational transitions, $\Delta J = \pm 2$. The amount of internal energy transferred to translation in a V-T, R transfer of this kind will be

$$\Delta E_{\text{int}} = \Delta E_{\text{vib}} \pm \Delta E_{\text{rot}}.$$

If the rotational transition accompanying a $v = 1 \rightarrow 0$ vibrational transition is $\Delta J = \pm 2$, the energy-level diagram for the relaxation of a state, (v=1) (J*) will be as shown in Fig. 3.13. The most probable transition is likely to be that involving the smallest transfer to translation, i.e. from (v=1) (J*) to (v=0) (J*+2). For this transition,

$$\Delta E_{\text{int}} = \Delta E_{\text{vib}} - \Delta E_{\text{rot}} = h\nu - B(4J^* + 6), \tag{3.13}$$

so that the magnitude of the effect of the simultaneous rotational transition on the total relaxation process will depend critically on $B = \dfrac{h}{8\pi^2 c I_B}$, the rotational constant of the molecule. The slow V-(T, R) transfer into the (v=0) (J^*+2) level will be followed by very rapid thermal equilibration of the rotational levels in the (v=0) state: i.e., if J^* is the most highly populated rotational level at the experimental temperature, there will be rapid R-T transfer, predominantly from (J^*+2) to J^*. The total effective transfer from internal to translational energy, as measured by acoustic or shock-tube experiments, will thus be ΔE_{vib}, but the rate-controlling step will be the slower V-(T, R) transfer involving the smaller quantity ΔE_{int}.

We will now speculate on what happens if we apply this model to the Lambert-Salter plot. This is illustrated in Fig. 3.14. Here the ordinate is ΔE, the abscissa $\log_{10} Z_{10}$, and the diagonal line represents the relation (3.12) with the 'true' gradient C corresponding to *actual* V-T transfer. The plotted point α

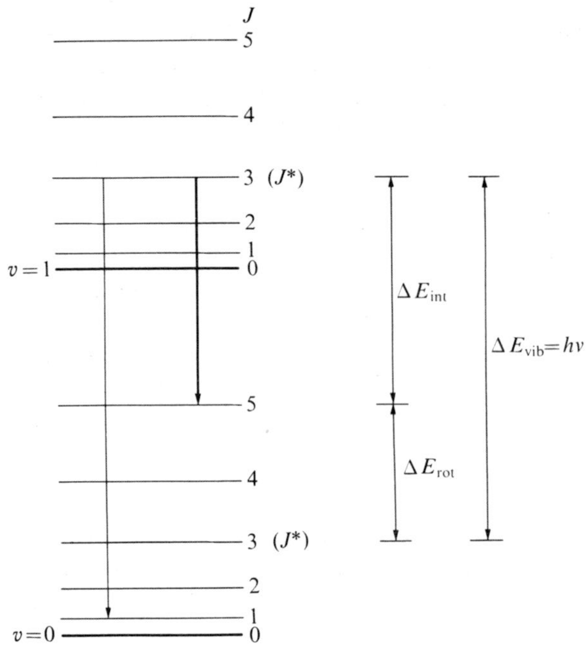

FIG. 3.13. Energy-level diagram for simultaneous vibrational and rotational transitions[166]

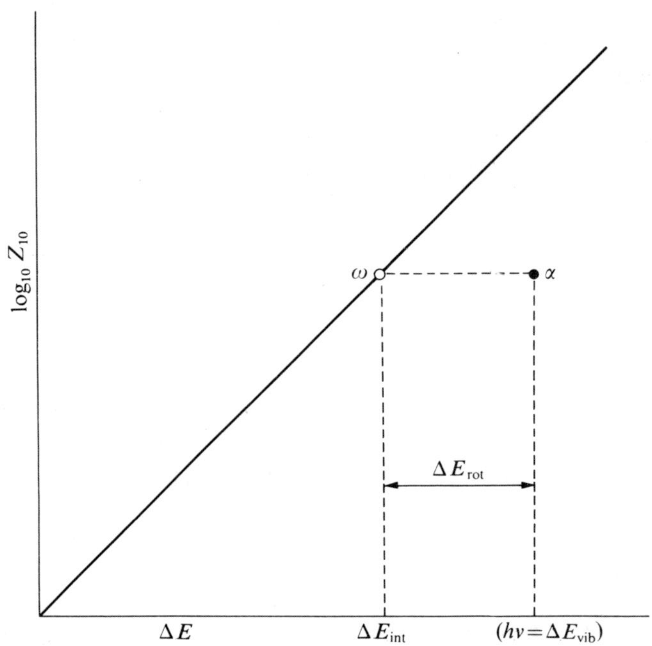

FIG. 3.14. Effect of simultaneous vibrational and rotational transitions on a Lambert-Salter plot[166]

represents a molecule for which $h\nu = \Delta E_{\text{vib}}$. If ΔE_{rot} is the rotational contribution to the V-(T, R) transfer, the actual size of this transfer will be

$$\Delta E_{\text{int}} = h\nu - \Delta E_{\text{rot}}, \tag{3.14}$$

corresponding to the point ω which lies on the diagonal line. The effect of simultaneous rotational transitions is thus to shift experimental points on a Lambert-Salter plot *horizontally to the right.* The magnitude of the shift will be ΔE_{rot}, which is dependent on the rotational constant B, and will be most marked for molecules with small moments of inertia, such as some of the hydrides.

The labelled experimental points on the plot for hydrogen-containing molecules in Fig. 3.10 show systematic deviations which are in accord with this view. They fall into three groups: (a) the polyhydrides CH_4, PH_3, AsH_3, SiH_4 all fall to the *right* of line I; (b) the symmetric-top molecules CH_3Cl, CH_3Br, CH_3I together with C_2H_4 all fall close to the line; (c) the ring molecules c-C_3H_6, C_2H_4O, and the linear molecule C_2H_2 all fall to the *left* of the line. These groups are differentiated by moments of inertia and rotational constants, as shown in Table 3.9. Group (a) show large values of A, B, and C so that both J and K rotational levels will be widely spaced: group (b) show large values of A and much smaller values of B and C, so that only K levels will be widely spaced: group (c) show small values of A, B, and C, so that all rotational level spacing will be small.

TABLE 3.9

Moments of inertia ($I/10^{-40}$ g cm^2) *and rotational constants* (A, B, C/cm^{-1}) (*ref. 166*)

Molecule	I_A	A	I_B	B	I_C	C
			GROUP (*a*)			
CH_4	5·3	5·3	5·3	5·3	5·3	5·3
PH_3	4·4	6·4	6·2	4·5	6·2	4·5
AsH_3	4·4	6·4	6·7	4·2	6·7	4·2
SiH_4	9·5	2·9	9·5	2·9	9·5	2·9
			GROUP (*b*)			
C_2H_4	5·8	4·9	28	1·0	34	0·83
CH_3Cl	5·5	5·1	57	0·49	57	0·49
CH_3Br	5·5	5·1	90	0·31	90	0·31
CH_3I	5·5	5·1	100	0·28	100	0·28
			GROUP (*c*)			
c-C_3H_6			42	0·67	42	0·67
C_2H_4O	32	0·88	39	0·71	59	0·47
C_2H_2			23·8	1·2		

Nothing approaching an accurate calculation of the values of ΔE_{rot} can be made for polyatomic molecules, but a rough estimate of the expected order of magnitude of the effect can be made by analogy with a diatomic molecule with $B = 5\ \text{cm}^{-1}$. For such a molecule J^* will be 4 at 300 K, and by eqn (3.13) $\Delta E_{rot} = 110\ \text{cm}^{-1}$ for $\Delta J = 2$. There are three points on Fig. 3.10 corresponding to molecules from each of the three groups, which show identical values of $\log_{10} Z_{10} = 2{\cdot}78$: AsH_3 ($\nu_{min} = 905\ \text{cm}^{-1}$), CH_3Cl ($\nu_{min} = 732$), and C_2H_4O ($\nu_{min} = 685$). Identical values of Z_{10} correspond to identical values of ΔE_{int}, so that the horizontal displacements between the three points should correspond to eqn (3.14) and

$$\Delta E_{rot}\ (AsH_3) - \Delta E_{rot}\ (CH_3Cl) = 905 - 732 \sim 170\ \text{cm}^{-1}$$
$$\Delta E_{rot}\ (CH_3Cl) - \Delta E_{rot}\ (C_2H_4O) = 732 - 685 \sim 50\ \text{cm}^{-1}.$$

Both displacements are of the same order of magnitude as the $110\ \text{cm}^{-1}$ calculated for the $\Delta J = 2$ transition for a diatomic molecule with $B = 5\ \text{cm}^{-1}$. If we assume that in the polyatomic molecules *both* J and K levels decrease by 2, reference to the rotational constants in Table 3.9 shows that this will be the kind of effect predicted. Such an effect is clearly much too small to account for the large separation of lines I and II, and it will be unimportant for the non-hydrogen-containing molecules, which all have large moments of inertia, so that ΔE_{rot} becomes small enough to be negligible. The same will apply to the more complex hydrogen-containing molecules which lie below CH_3I on the plot in Fig. 3.10.

3.5. Mixtures with one relaxing component

3.5.1. Relaxation in mixtures

If a relaxing gas, A, is mixed with a non-relaxing gas, B, such as helium, there are two collision processes in which V-T energy transfer may occur:

$$(1)\quad A^* + A \rightarrow A + A \quad \text{(V-T)}$$
$$(2)\quad A^* + B \rightarrow A + B \quad \text{(V-T)}$$

Since (1) and (2) will have different collisional efficiencies, the result will be a composite relaxation time for A*, given by

$$\frac{1}{\beta_A} = \frac{1-x}{\beta_{AA}} + \frac{x}{\beta_{AB}}, \tag{3.15}$$

where x is the mole fraction of B in the mixture, β_{AA} is characteristic of process (1), and β_{AB} of process (2). This may be represented graphically by a *linear* plot of reciprocal relaxation time $1/\beta$ against the mole fraction x. Such a plot for experimental measurements made with CF_4 + He mixtures[193] is shown in Fig. 3.15. The intercept on the $x = 1{\cdot}0$ axis gives a value for β_{AB} from which Z_{AB} may be calculated by the procedure given in § 1.2.3. Z_{AA}, which is simply Z_{10} for the pure gas A, is calculated from the intercept on the $x = 0$ axis.

The relaxation of mixtures of *two relaxing* gases is always dominated by intermolecular V-V transfer, and will be considered in Chapter 4. The only

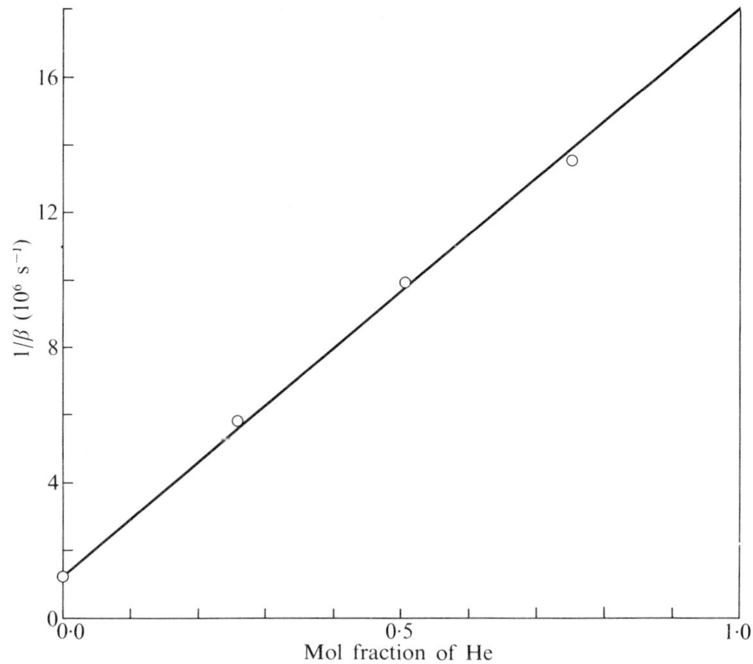

FIG. 3.15. Reciprocal relaxation times for CF_4 + He mixtures (data from Olson and Legvold[193])

additives we shall consider here are inert gases or diatomic or triatomic molecules with such high vibrational frequencies that easy, near-resonant V-V transfer is ruled out. For such mixtures, the relation between Z_{AA} and Z_{BB} depends, in the same way as most other physical properties, on the intermolecular forces acting between A/A and A/B. Mixtures will thus fall into three classes: (a) those where similar forces act between A/A and A/B; (b) those where there are specific forces between A/A which are not operative for A/B; (c) those where there are specific forces between A/B which are not operative for A/A. There is a very large mass of published information about vibrational relaxation in mixtures, and we shall select cases which shed interesting light on the characteristic properties and differences of the three classes.

3.5.2. Mixtures without specific interactions

The relaxation of mixtures of non-polar or weakly polar gases with inert gases, or non-polar diatomic gases can be handled qualitatively in terms of SSH theory. Quantitative predictions show varying success owing to the difficulty of obtaining satisfactory collision parameters. The most important variable in comparing a series of different inert additives is the reduced mass of the collision, $\mu = m_A m_B/(m_A + m_B)$, and simplified theory (eqn 3.3) would predict an exponential dependence of Z_{10} on $\mu^{\frac{1}{3}}$. Table 3.10 shows values of Z_{AB} for some

TABLE 3.10.

Collision numbers for V-T transfer in mixtures at 300 K

Mixture	ν_{min} (cm^{-1})	μ	Z_{AB}	Reference
$O_2 + O_2$	1554	16	1·03(8)	197
+ He		3·6	3·84(5)	
+ D_2		3·6	5·74(5)	
+ H_2		1·9	3·36(4)	
$CO_2 + CO_2$	673	22	5·3(4)	180
+ He		3·7	3·2(3)	198
$N_2O + N_2O$	589	22	5·5(3)	200
+ He		3·7	1·6(3)	200
$CF_4 + CF_4$	437	44	2·22(3)	193
+ Ar		27·5	3·73(3)	
+ Ne		16·4	8·59(2)	
+ He		3·8	2·23(2)	
$CHF_3 + CHF_3$	507	35	2·18(3)	193
+ Ar		28·4	7·43(3)	
+ Ne		15·7	2·43(3)	
+ He		3·8	2·98(2)	
$CF_2Cl_2 + CF_2Cl_2$	261	60·5	1·09(2)	193
+ Ar		30·0	3·38(2)	
+ Ne		17·3	1·12(2)	
+ He		3·9	7·2(1)	
c-C_3H_6 + c-C_3H_6	740	21	1·27(3)	199
+ Ar		20·5	2·27(3)	
+ Ne		13·5	1·68(3)	
+ He		3·7	5·40(2)	

typical mixtures. All show roughly the same pattern of behaviour: the effectiveness of inert gas additives decreases steadily with increasing mass: collisions with He are always more effective than self-collisions. For the polyatomic molecules collisions with Ar are always less effective than self-collisions, in spite of smaller reduced mass. Collisions of O_2 with D_2 and H_2 seem to follow the same pattern, but these must be regarded with some reserve, owing to the possibility of intermolecular V-R transfer. Temperature-dependence has been investigated for only two of the systems in Table 3.10, and both the CO_2[198] and O_2[203] mixtures show the normal Landau-Teller $T^{\frac{1}{3}}$ temperature-dependence. Relaxation times for mixtures of CO with H_2, He, Ne, and Kr have been measured over a large temperature range by Millikan,[201,202] and show normal temperature-dependence at higher temperatures. Fluorescence measurements at $T < 300$ K for CO + He and CO + H_2 mixtures show the same kind of deviation at lower temperatures as pure CO[202] (cf. Fig. 3.3).

3.5.3. Mixtures showing anti-catalytic behaviour

These are mixtures where the relaxation behaviour of component A is dominated by special factors characteristic of A + A collisions which do not operate for A + B collisions. A striking example is provided by mixtures of inert gases with HCl, where in the pure gas strong intermolecular attractive forces due to dipole-dipole interaction, together with some contribution from V-R transfer due to the low moment of inertia, lead to highly efficient transfer and to an inversion of temperature-dependence. A Landau-Teller plot of experimental values of Z_{AB} for mixtures of HCl with He and with Ar is shown in Fig. 3.16. The plot for HCl + HCl collisions, taken from Fig. 3.7, is shown

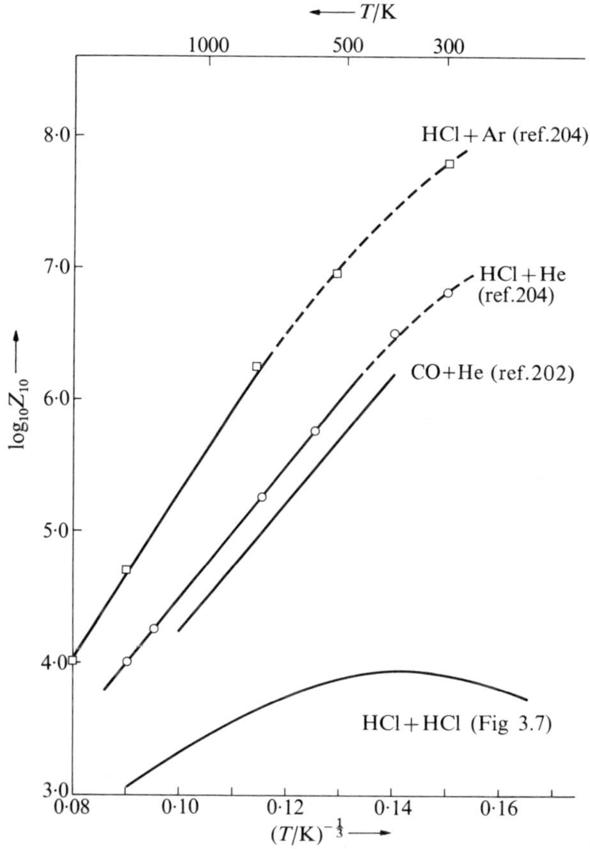

FIG. 3.16. Landau-Teller plots for HCl mixtures

for comparison, also a plot for CO + He collisions, as an example involving a non-polar diatomic molecule of similar mass and frequency (M(HCl) = 36·5: M(CO) = 28: ν(HCl) = 2886 cm^{-1}: ν(CO) = 2143 cm^{-1}). It will be seen that the special characteristics of HCl + HCl collisions, very high efficiency and inversion of temperature-dependence, are entirely absent from the collisions with inert gases, which resemble collisions between CO and He (described in § 3.5.2). The HCl + He collisions are slightly less efficient than the CO + He collisions, which can be attributed to the lower mass and vibration frequency of CO, and show the same temperature-dependence. The HCl + Ar collisions are less efficient, corresponding to the larger reduced mass. Other strongly polar gases, both diatomic and polyatomic, show similar behaviour in mixture with inert gases. Mixtures of HF with He and Ar,[122] of HBr with He,[64] and of SO_2 with Ar[205] all relax much more slowly than the pure gases and show normal temperature-dependence. Mixtures of H_2O with He, Ar, and N_2 relax more slowly than pure H_2O, but show very little temperature-dependence.[232] A case which is difficult to explain is BCl_3, which has recently been investigated by infrared double-resonance techniques.[209] This is a non-polar, planar molecule: the lowest vibrational frequency is $\nu_4 = 243$ cm^{-1}, and for the pure gas at 300 K, $Z_{10} = 145$ for relaxation of the whole vibrational energy content. This is slightly larger than predicted by the Lambert-Salter plot, which suggests that relaxation occurs by a normal repulsive interaction. But collisions with He, $Z_{AB} = 1860$, are far less efficient, although collisions with the other inert gases, Ne ($Z_{AB} = 1860$), Ar ($Z_{AB} = 4035$), Kr ($Z_{AB} = 4440$), Xe ($Z_{AB} = 5870$) follow the usual mass dependence relative to He.

3.5.4. *Mixtures showing catalytic behaviour*

These are mixtures whose relaxation behaviour is dominated by special characteristics of A + B collisions, which do not operate for self-collisions. The classical case is $CO_2^* + H_2O$, first investigated by Eucken in 1937.[210] Pure CO_2 ($\nu_{min} = 673$ cm^{-1}) conforms to the Lambert-Salter plot (Fig. 3.12), showing very slow relaxation, $Z_{10} \sim 50\,000$ at 300 K, and normal temperature-dependence. For mixtures of $CO_2^* + H_2O$ recent investigations show a value of $Z_{10} = 26$ at 300 K, *rising* steadily with increasing temperature to 29 at 468 K.[211] There are no near-resonant vibrational modes in the two molecules, which might lead to rapid V-V transfer, and the striking catalytic effect with reversed temperature-dependence must be attributed to incipient chemical reaction between CO_2 and H_2O (which can form H_2CO_3).[212,213] This is discussed in detail by Widom and Bauer,[213] who point out that the charge distribution of the CO_2 molecules is strongly positive on the central C-atom and negative on the two O-atoms. The formation of H_2CO_3 is likely to involve the initial steps

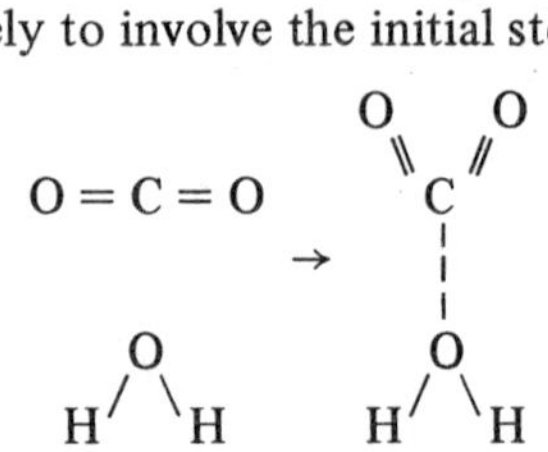

This will correspond to strongly oriented attractive interaction, with a potential well depth ~8 kcal mol^{-1} which, by analogy with the behaviour of pure polar

gases discussed earlier, is quite sufficient to account for the phenomena. CO_2^* is also very rapidly relaxed by HF, $Z_{AB} = 160$ at 300 K,[259] with negative temperature-dependence, and this has been interpreted by Shin[260] in terms of the formation of a hydrogen-bonded dimer,

$$F{-}H\cdots\cdots O{=}C{=}O,$$

whose heat of formation is estimated as between 2 and 4 kcal mol^{-1}. Similar but less marked effects are shown by $N_2O^* + H_2O$ mixtures.[210] Pure N_2O ($\nu_{min} = 589$ cm^{-1}) shows a value of $Z_{10} = 7500$ at 300 K, with normal temperature-dependence. Mixtures of N_2O^* with H_2O show $Z_{AB} \sim 105$ at 300 K rising to 210 at 473 K. There is no incipient chemical reaction here, but some form of oriented quadrupole-dipole interaction would produce the observed effects. The most striking effect of all is shown by $HF^* + H_2O$ and $HF^* + D_2O$ mixtures.[214] The strongly polar HF shows $Z_{10} = 57$ in the pure gas at 300 K (cf. § 3.3.2). Mixtures with H_2O and D_2O show $Z_{AB} \sim 1{\cdot}5$: i.e. the transfer occurs at almost every gas kinetic collision. Neither V-V nor V-R transfer could account for such efficient relaxation of HF ($\nu = 4138$ cm^{-1}), and Hancock and Green[214] attribute this to the formation of an oriented dimer,

H\
  O······H–F,
H/

whose heat of formation is estimated theoretically as 10 kcal mol^{-1} (higher than the heat of dimerization of pure HF, 6 kcal mol^{-1}).

Collisions of molecules with *free atoms* also show marked catalytic effects. Measurements obtained by laser fluorescence at 300 K and by shock-tube techniques at higher temperatures are shown in Table 3.11. Relaxation times, rather than collision numbers, are given owing to the absence of reliable collision diameters for such mixtures. Data for self-collisions and for collisions with inert gas molecules are included for comparison. The data must be treated with some reserve, as the detailed analysis of the experimental results is not always simple, but it is clear that O, Fe, F, Cl, and Br atoms are very efficient energy-transfer catalysts, and H atoms rather less efficient. Two alternative explanations have been advanced:

(a) All the atoms quoted, except H, possess resultant orbital angular momentum, which renders possible a vibronic transition, as was proposed by Nikitin for NO.[146,229] The specially high efficiency for $HF^* + Br$ has been explained by Quigley and Wolga[234] on the basis of near-resonant E-V transfer between the spin-orbit splitting in Br($^2P_{\frac{1}{2}} \leftrightarrow {}^2P_{\frac{3}{4}}$: $\Delta E = -3685$ cm^{-1}) and the vibrational relaxation of HF^* ($v{=}1$). A similar effect has been predicted for $H_2^* + Br$.[242]

(b) Very strong attractive forces would be expected to operate between free atoms and molecules, and these (or perhaps incipient chemical reaction, as between CO_2 and H_2O) are responsible for the efficient transition. Classical calculations for $O_2 + O$ collisions by Brieg[225] and of $HF + F$ by Thompson[226] and by Thommarsen and Berend[230] along these lines show fair agreement with

TABLE 3.11

V-T relaxation times for mixtures with atoms

Mixture	Temp. (K)	β (μs atm)	Reference
$N_2^* + N_2$	300	10^6	51
+ He	300	6900	51
+ O	300	11·0	217
$+ N_2$	3000	50·0	239
+ O	3000	1·8	215
CO* + CO	2000	80	48
+ Ar	2000	500	222
+ O	2000	0·04	216, 219
+ Fe	2000	0·06	222
$O_2^* + O_2$	1500	20	101
+ Ar	1500	100	240
+ O	1500	0·03	218
$CO_2^* (\nu_3) + CO_2$	300	(V-V) 3·8	241
+ Ne	300	23	45
+ O	300	0·21	221
+ O	300	0·20	233
+ Cl	300	0·003	233
+ F	300	0·013	233
+ H	300	0·39	233
HF* + HF	300	0·022	126
+ Ar	300	> 22	214
+ O	300	0·014	224
+ F	300	0·15	224
+ Br	300	0·0014	234
+ H	300	19	235
HCl* + HCl	300	1·9	129
+ Ne	300	1500	236
+ Ar	300	12000	236
+ O	300	0·025	237
+ Cl	300	0·0047	238

All relaxation times are for M* (ν=1)

experiment. The efficiency of H atoms, which are not endowed with orbital angular momentum, can be explained *only* in this way.

There is not enough experimental evidence to decide between the two alternatives (a) and (b), and probably both effects can operate simultaneously. The few measurements of temperature-dependence, which is either positive (Landau-Teller), or very weak, are inconclusive. For some of the mixtures in Table 3.11, e.g. O_2^* + O and HCl* + Cl, there is the possibility of an actual chemical exchange reaction being responsible for the relaxation:

$$\text{O–O}\ (v=1) + \text{O} \leftrightarrow \text{O} + \text{O–O}\ (v=0),$$

or

$$\text{Cl} + \text{H–Cl}\ (v=1) \leftrightarrow \text{Cl–H}\ (v=0) + \text{Cl}.$$

These reactions would be expected to show the observed temperature dependence. Calculations made on this basis for $O_2^* + O$ by Bauer and Tsang[244] agree with the experimental relaxation rates[218] within a factor of $\times 2$. A very detailed theoretical calculation for the systems $HCl^* + H$ and $HCl^* + Cl$, based on LEPS potential energy surfaces, has been published by Wilkins,[243] and also shows fair agreement with experiment. Experimental evidence is insufficient to decide whether chemical reaction has actually occurred or not, but it is quite clear that even *potentially* 'reactive' collisions lead to highly efficient V-T transfer.

A different kind of catalytic behaviour can arise in collisions with hydride additives, involving a vibrational transition in the substrate simultaneous with a rotational transition in the hydride additive. This usually seems to involve a near-resonance between the $v=1 \rightarrow 0$ transition of a low-frequency vibrational mode of the substrate and a $\Delta J = 1, 2, 3$, or 4 rotational transition in the additive. The most convincing evidence for this mechanism is afforded by hydrogen mixtures, where differences in the efficiency of o-H_2 and p-H_2 can be ascribed only to rotational effects. For $CO_2^* + H_2$ mixtures there is near-resonance between the $CO_2\ (\nu_2)\ v=1 \rightarrow 0$ transition and $J=1 \rightarrow 3$ transition of o-H_2:

$$CO_2\,(010) + o\text{-}H_2\,(J=1) \rightarrow CO_2\,(000) + o\text{-}H_2\,(J=3) + 80{\cdot}3\ \mathrm{cm^{-1}}$$

but not with the $J=2 \rightarrow 4$ transition of p-H_2:

$$CO_2\,(010) + p\text{-}H_2\,(J=2) \rightarrow CO_2\,(000) + p\text{-}H_2\,(J=4) - 146\ \mathrm{cm^{-1}},$$

and acoustic measurements at 300 K show[253]

$CO_2^* + CO_2$	Z_{10}	$= 41900$
$CO_2^* + p$-H_2	Z_{AB}	$= 488$
$CO_2^* + o$-H_2	Z_{AB}	$= 227.$

Similar differences are shown by shock-tube experiments at higher temperatures.[254] In contrast, for $CO + H_2$ mixtures vibrational relaxation is two times faster for p-H_2 than for o-H_2,[255] and here the process

$$CO\,(v=1) + p\text{-}H_2\,(J=2) \rightarrow CO\,(v=0) + p\text{-}H_2\,(J=6) + 88\ \mathrm{cm^{-1}},$$

is more nearly resonant than the alternative

$$CO\,(v=1) + o\text{-}H_2\,(J=3) \rightarrow CO\,(v=0) + o\text{-}H_2\,(J=7) - 327\ \mathrm{cm^{-1}}.$$

The experimental results, which were obtained by fluorescence quenching, are in fair agreement with theoretical calculations based on this model.[256] In both mixtures the simple V-T mass effect and the simple semiclassical V-R mechanism of § 3.2.3 may also contribute to the fast relaxation, but neither can explain the observed differences. According to the simple mass effect there should be no difference between the efficiencies of o-H_2 and p-H_2, while the semiclassical V-R model predicts o-H_2 to the more efficient for both mixtures.

A similar explanation has been applied to BCl_3^* + HCl mixtures. BCl_3 (ν_4 = 243 cm^{-1}), whose apparently anomalous *anti*-catalytic behaviour in mixtures with He was discussed in § 3.5.3, shows more efficient relaxation in collision with HCl (Z_{AB} = 91) than in self-collision (Z_{10} = 145). This can be attributed to near-resonant J=4→6 or J=3→5 rotational transitions in HCl and the v=1→0 vibrational transition of BCl_3,[196] which will have a high probability in collisions at 300 K, where the J=3 and J=4 levels of HCl lie near the peak of the Boltzmann distribution of rotational levels. DCl is less efficient (Z_{AB} = 146); here near-resonance with BCl_3 (ν_4) would only occur with the J=9→11 and J=10→12 transitions of DCl, where the initial J-states have very low populations at 300 K. The temperature-dependence will thus involve Boltzmann population of the relevant rotational levels in addition to the transition probability. The rapid relaxation of Cl_2 in collisions with HCl, Z_{10}/Z_{AB} ~100 at 400 K, may also be attributed to some kind of V-R transition; here DCl is again less efficient[257] than HCl by a factor of 2 to 4. V-R transfer has been invoked by Shields[258] as an alternative mechanism for the catalytic behaviour of CO_2 + H_2O mixtures (discussed above): the observation that D_2O is only $\frac{1}{3}$ as efficient as H_2O would support this, and both chemical and V-R effects may be operating.

The molecular beam technique has opened up the possibility of investigating relaxation in collisions between molecules and *ions*, where another kind of specific interaction can occur. Udseth *et al.*[231] have investigated the transition probability for vibration relaxation of H_2, D_2, and HD in collision with H^+, and find that this is independent of isotopic composition, i.e. $Z_{H^+H_2} \sim Z_{H^+HD} \sim Z_{H^+D_2}$. This cannot be explained in terms of any of the physical models so far discussed for V-T or V-R transfer. They suggest that the interaction which leads to vibrational relaxation is the effect of the charged proton on the *electrons of the molecular bond.* This experiences a time-dependent 'dilution' as the proton moves past, leading to efficient vibrational relaxation.

3.6 The information theoretic approach

A new method for treating the final energy state distribution for the products of chemical reaction has recently been described by Levine and Bernstein.[245] This method is based on a comparison of the actual final state distribution $P(E_j)$ to a calculated *a priori* distribution $P^0(E_j)$, which is derived on the assumption that all quantum states at a given total energy distribution are equally probable. The *entropy deficiency* which is equal to the *information content* of the *actual* distribution is given by[227]

$$\Delta S = R \ln (P^0(E_j)/P(E_j)).$$

The comparison may also be expressed as a *surprisal*, which is simply a measure of the deviation of the actual probability of the final distribution from that of the *a priori* distribution:[245]

$$I(E_j) = -\ln (P(E_j)/P^0(E_j)),$$

so that

$$P(E_j) = P^0(E_j) \exp[-I(E_j)] .$$

This concept has been applied to non-reactive inelastic collisions,

$$AB^* (v) + M \rightarrow AB^* (v') + M,$$

by Rubison and Steinfeld[246] and by Procaccia and Levine.[247,248] The find that the necessary and sufficient condition for the experimentally observed exponential relaxation of the mean vibrational energy of AB* (assumed to be a harmonic oscillator) is that the detailed rate constants for each transition, $k(v \rightarrow v')$ satisfy

$$\sum_{v'} (v - v)\, k(v \rightarrow v') = \alpha\, [\langle\!\langle v(\mathrm{eq})\rangle\!\rangle - v] ,$$

where α is the bulk vibrational relaxation rate in a heat bath, and $\langle\!\langle v(\mathrm{eq})\rangle\!\rangle$ is the equilibrium, bulk, average vibrational state at the bath temperature T. Information theory, applying the principle of minimizing the information content (maximizing the entropy), subject to this constraint, then provides the following functional form for the rates:

$$k(v \rightarrow v') = Rk^0 (v \rightarrow v') \exp [-\lambda_v |v - v'| (h\nu/kT)] , \qquad (3.16)$$

where $k^0 (v \rightarrow v')$ is a 'prior' relative rate constant for vibrational energy transfer. R is a (v-independent) collision rate and λ_v determines the 'spread' of the final vibrational states v' in a single collision. This procedure has been applied[248] to the relaxation of HF from a state of vibrational disequilibrium, and gives good agreement with experimental data, obtained by analysis of laser fluorescence experiments, for the relaxation of HF ($v = 2, 3, 4$).[249] Good agreement is also shown with the results of classical trajectory computations[252] for the process,

$$Cl + Cl_2(v) \rightarrow Cl_2 (v') + Cl.$$

The procedure implies that, for a given diatomic molecule, λ_v is essentially determined entirely by the temperature and is constant for all relaxation channels. Bernstein[250] has pointed out that a similar exponential expression arises for *rotational* energy transitions between states J and J'. This is in striking agreement with an analysis of experimental data on the chemiluminescence of rotationally excited HCl ($J \leqslant 16$) by Polanyi and Woodall,[67] who found that the overall downward relaxation process is best described in terms of a relation

$$P_{J \rightarrow J'} = N \exp(-C\Delta E), \qquad (3.17)$$

where $P_{J \rightarrow J'}$ is the collision probability of the transfer, ΔE is the energy difference between the two states J and J', and N and C are constants for a particular temperature.

It is interesting that eqns (3.16) and (3.17) are closely analogous with the Lambert-Salter relation (3.12), which applies to $v = 1 \rightarrow 0$ transitions of the lowest vibrational mode for a wide variety of polyatomic molecules at 300 K. It is also a striking coincidence that Polanyi and Woodall's estimated value of $C = 8{\cdot}5 \times$

10^{-3} cm for rotational transitions in $HCl^* + H_2$ collisions at 230 K is almost equal to the Lambert-Salter value of $C_I = 8{\cdot}3 \times 10^{-3}$ cm for vibrational transitions in hydrogen-containing molecules at 300 K (cf. § 3.4.2). Eqn (3.16) predicts an exponential factor, equivalent to C in eqns (3.12) and (3.17), varying inversely with temperature. If the molecules which conform to the Lambert-Salter plot follow the normal Landau-Teller temperature-dependence, as inferred in § 3.4.2, this would predict that the constant C_I should vary inversely with $T^{\frac{1}{3}}$, showing qualitative, but not quantitative agreement. It would be premature to suggest that information theory can be applied in its present form to explain the Lambert-Salter plot, but this looks a promising line of approach. A further difficulty is that the Lambert-Salter plot does not fit the available data for $v=1{\rightarrow}0$ transitions in diatomic molecules. The relation between the Polanyi-Woodall equation for rotational transitions and information theory will be further discussed in § 5.5.

4
VIBRATION–VIBRATION TRANSFER

4.1. The SSH theory for non-resonant and resonant transfer

The SSH expression (3.2) for $P_{k-l(\text{b})}^{i-j(\text{a})}$, the probability that a vibrational mode (a) of one molecule will change its quantum state from i to j, while simultaneously a second mode (b) in the other molecule will change its state from k to l, applies equally to V-T transfer (where $k = l$) and to V-V transfer (where $k \neq l$). For the V-T case, ΔE, the energy exchanged between vibrational and translational degrees of freedom, is

$$\Delta E = h\nu_\text{a}(i-j).$$

For the V-V case

$$\Delta E = h\nu_\text{a}(i-j) + h\nu_\text{b}(k-l) \tag{4.1}$$

(cf. § 3.1.3). Non-resonant V-V transfers are conveniently handled in terms of the analytical solution of (3.2) due to Tanczos (eqn (3.3)), which was used as the basis for discussion of V-T transfer in Chapter 3, and is valid for values of $\Delta E \geqslant 200\ \text{cm}^{-1}$. Similar criteria should apply for dependence on vibrational frequency, quantum number, single *versus* multiple quantum transfer, collision mass, temperature, and intermolecular potential.

For a V-V transfer between different molecular species the additional possibility arises of a *resonant* transition, where

$$h\nu_\text{a}(i-j) = h\nu_\text{b}(k-l),$$

and $\Delta E = 0$. For resonant and near-resonant transfers a different analytical solution, also due to Tanczos,[27] applies for values of $\Delta E \leqslant 50\ \text{cm}^{-1}$:

$$P_{k-l(\text{b})}^{i-j(\text{a})} = P_0(\text{a})\,P_0(\text{b})\,[V^{i-j}(\text{a})\,.\,V^{k-l}(\text{b})]^2\,\frac{64\pi^2\mu kT}{\alpha^2 h^2}\,\exp\,[\epsilon/kT]. \tag{4.2}$$

Resonant transfers are much faster than non-resonant transfers, and there is no term in ΔE, so that all near-resonant transfers with $\Delta E \leqslant 50\ \text{cm}^{-1}$ behave as though resonant. Dependence on the frequency of the exchanging modes arises through the pre-exponential term $[V^{i-j}(\text{a})\,.\,V^{k-l}(\text{b})]^2$. Since each squared vibration factor for a single-quantum transition, $[V^{(i-l)-i}]^2$, is inversely proportional to frequency from eqn (3.5), this gives rise to inverse dependence of P on the square of the frequency. Resonant transfers between high frequency modes will thus have a collision probability much smaller than unity. This is because, although there is no net transfer of energy from vibration to translation, the change in vibrational quantum state can only be brought about by a time-dependent force.[5] (The same factor also operates for non-resonant transfers (eqn (3.3)), but will usually be far outweighed by the exponential dependence on ΔE.) The last pre-exponential term predicts a weak *positive* dependence on temperature and a weak *positive* dependence on reduced mass. The only exponential term is that depending on the intermolecular attractive parameter, $\exp[\epsilon/kT]$, so that quite a small value of ϵ will be sufficient to outweigh the pre-exponential term and produce an overall *negative* temperature dependence for resonant V-V transfer, which will in general be much more sensitive to intermolecular attractive forces than non-resonant transfer. Quantitative treatment of such transfers by SSH theory will again be unsatisfactory due to inadequacy of the assumed potential.

4.2. Intermode transfer in polyatomic gases

4.2.1. Single relaxation process

Intramolecular transfer of energy between different vibrational modes can only occur in collision, as the energy discrepancy between the modes must be made up as translational energy. For a molecule with two active vibrational modes of frequency ν_1 and ν_2 there are three possible vibrational transitions, which are illustrated on the energy-level diagram in Fig. 4.1:

(a) Transfer of translational energy into 0→1 excitation of the mode ν_1, with relaxation time β_1:

$$A^*\,\nu_1\,(v{=}1) + A \leftrightarrow A\,(v{=}0) + A + \Delta E.$$

(b) Transfer of translational energy to 0→1 excitation of the mode ν_2. with relaxation time β_2:

$$A^*\,\nu_2\,(v{=}1) + A \leftrightarrow A\,(v{=}0) + A + \Delta E.$$

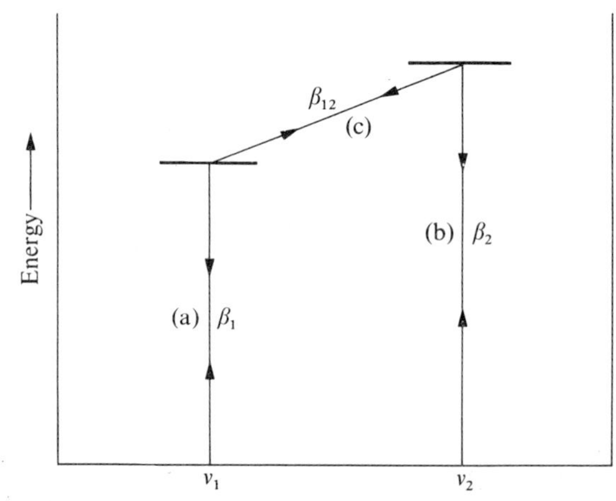

FIG. 4.1. Energy-level diagram showing transitions for a molecule with two active vibrational modes

(c) The *complex* transfer of one quantum of vibrational energy from mode ν_1, plus the necessary increment of translational energy to give 0→1 excitation of the mode ν_2, with relaxation time β_{12}:

$$A^* \nu_1 (v{=}1) + A \leftrightarrow A^* \nu_2 (v{=}1) + A + \Delta E.$$

The values of β_1, β_2, and β_{12} may, in principle, be calculated by SSH theory. Since only one molecular species is involved in all three processes, their relative magnitudes will be mainly determined by the values of ΔE. For the model represented in Fig. 4.1 the fastest relaxation process will be the complex transfer (c), and the slowest the V-T relaxation of mode ν_2 by process (b). This means that $\beta_2 > \beta_1 > \beta_{12}$. Vibrational energy thus enters the molecule *via* process (a), which is rate controlling, and rapidly flows in complex collisions *via* the faster process (c) to the upper mode. Process (b) is too slow to play any role. This is known as a 'series' mechanism and is characterized by a single overall relaxation time β, which can be related to β_1 by $\beta_1 = (C_1/C_s)\beta$, where C_1 is the heat capacity contribution due to mode ν_1 alone, and C_s the total vibrational heat capacity (cf. § 3.4.1).[153]

This kind of behaviour is shown by the vast majority of polyatomic molecules. The most usual pattern of fundamental vibration frequencies is such that the energy gaps between the upper modes are much less than the energy level of the lowest mode. The general picture is that rapid vibration-vibration transfer maintains continuous equilibrium of vibrational energy between the various fundamental modes of the molecule, and that the whole of this energy relaxes in a single vibration-translation transfer process *via* the lowest mode. Two kinds of experimental evidence give support to this model. Ultrasonic dispersion measurements show clearly that, for these molecules, the whole vibrational heat content is transferred with a *single* relaxation time. More recently laser fluorescence measurements, which can yield the relaxation times of *both* the lowest mode *and* selected upper modes in suitable molecules, confirm both the acoustically determined relaxation times for lowest modes and the more rapid relaxation of upper modes. It is interesting to consider in detail the relaxation processes in an actual polyatomic molecule, CH_4, where SSH calculations for V-T transfer are in fair agreement with acoustic measurements (see Fig. 3.8). Collision numbers calculated by SSH theory by Stretton[152,23] for representative transitions are shown on an energy-level diagram in Fig. 4.2. It is clear that each of the upper modes, ν_2, ν_1, ν_3 can transfer its energy into the lowest mode, ν_4, or its first harmonic, much faster than the V-T transfer of ν_4. The first harmonic of ν_4 will relax very rapidly to the fundamental by single-quantum *resonant* intramolecular V-V transfer:

$$CH_4(2\nu_4) + CH_4 \rightarrow CH_4(\nu_4) + CH_4(\nu_4).$$

The overall picture is thus of relaxation of the total vibrational energy via ν_4 with a single relaxation time. This is in accord with the ultrasonic data quoted in § 3.4.1, and the calculated value $Z_{10} = 12\,500$ for relaxation of ν_4 is in fair

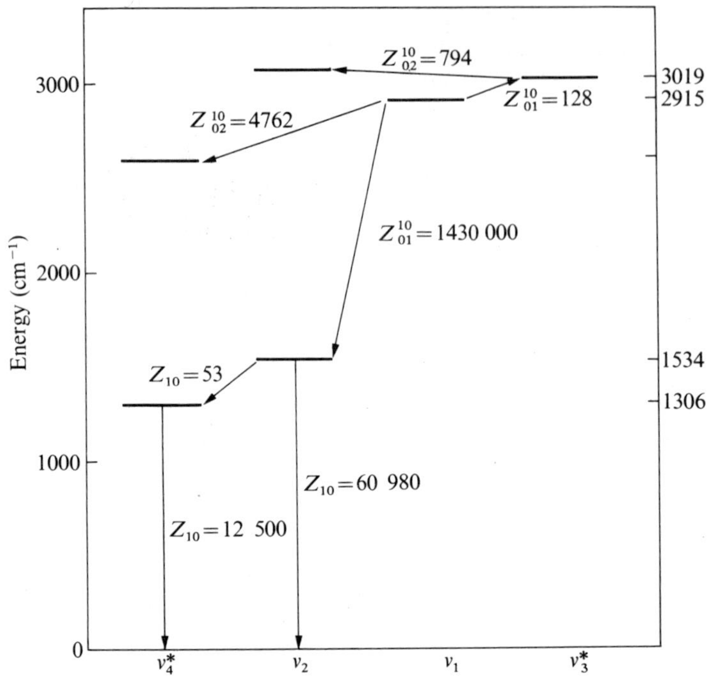

FIG. 4.2. Calculated collision numbers for V-V transitions in CH_4 (ref. 152)

agreement with the experimental value, $Z_{10} = 15\,160$ (Table 3.5). Laser fluorescence experiments were made on CH_4 by Yardley and Bradley Moore,[261] who excited the ν_3 mode by a mechanically chopped He-Ne laser beam, and obtained relaxation times for ν_4 and ν_3 by observing the phase shift between excitation and the fluorescence from both modes over a wide range of chopping frequencies. They found a value $Z_{10} = 14\,000$ for V-T relaxation of ν_4, in satisfactory agreement with the theoretical and ultrasonic values given above. ν_3 relaxed much more rapidly and they estimated that only 70 $CH_4 + CH_4$ collisions are required for this process. This is an order of magnitude faster than Stretton's calculated value for the rate-determining step:

$$CH_4(\nu_3) + CH_4 \rightarrow CH_4(2\nu_2) + CH_4 - 49 \text{ cm}^{-1},$$

shown on Fig. 3.8 ($Z_{02}^{10} = 794$). Yardley and Moore point out that there is an alternative rapid *exothermic* V-V transfer process:

$$CH_4(\nu_3) + CH_4 \rightarrow CH_4(\nu_2) + CH_4(\nu_4) + 180 \text{ cm}^{-1},$$

for which they calculate (using a slightly different SSH procedure) a collision number of 150. The SSH theory is thus qualitatively correct in predicting the general picture, but quantitative prediction for individual V-V transfer processes is not so reliable.

More extensive laser fluorescence measurements have been made by Flynn and collaborators on the two methyl halides, CH_3F and CH_3Cl, for which ultrasonic measurements also show single vibrational relaxation times (Table 3.5).[262,263] The data obtained from fluorescence are summarized on energy-level diagrams in Figs 4.3 and 4.4. The ν_3 mode of CH_3F and the ν_6 mode of CH_3Cl were pumped by pulses from a powerful *Q*-switched CO_2 laser. The rates of collisional V-V transfer to the other infrared active modes were obtained by measuring their fluorescence *risetimes.* Values of Z_r, the risetime collision number, calculated from the collision probabilities given in refs. 262 and 263, are shown on the figures. The subsequent fluorescence *decay rates* were measured, and the *same* (much lower) rate was found for *all* fluorescent states in the molecule. The inference follows that the whole vibrational manifold equilibrates rapidly by inter-mode V-V transfer (involving a multiplicity of paths) and this equilibrium is maintained while the total energy decays *via* one mode with a single relaxation time. This confirms the ultrasonic observations, and the overall fluorescent decay relaxation times, 175 ns for CH_3Cl and 2230 ns for CH_3F, are in fair agreement with the ultrasonic relaxation times, 204 ns for CH_3Cl and 1900 ns for CH_3F. The values of Z_{10} shown on the figures are calculated from the overall fluorescent relaxation times on the assumption

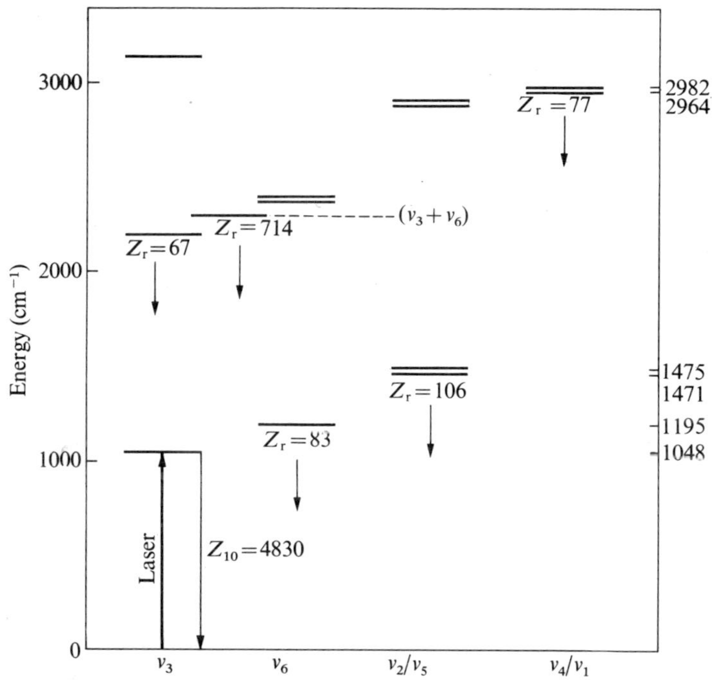

FIG. 4.3. Collision numbers for activation of infrared active modes of CH_3F (ref. 262). Z_r, calculated from fluorescence risetime; Z_{10}, calculated from fluorescence decay

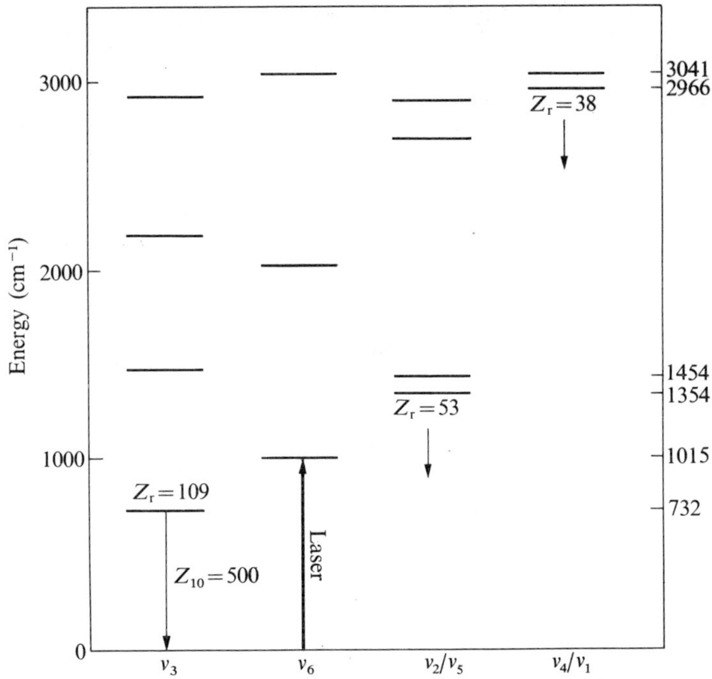

FIG. 4.4. Collision numbers for activation and deactivation of infrared active modes of CH_3Cl (ref. 263). Z_r, calculated from fluorescence risetime; Z_{10}, calculated from fluorescence decay

that relaxation occurs *via* the lowest mode ν_3. This is the carbon-halogen stretching frequency and has a very low vibrational amplitude factor compared to the other modes, which all involve the motion of hydrogen atoms (see Table 3.1 for CH_3Cl). SSH calculations by Stretton[152] predict that for CH_3F (also for CD_3Cl) the much larger vibration factor of the ν_6 mode outweighs the effect of its substantially higher frequency, so that the faster effective V-T relaxation should occur *via* the (*second* lowest) ν_6 mode; in contrast to CH_3Cl where ν_3 still relaxes faster. (Another simple molecule where the second lowest mode is predicted to relax faster is C_2D_2.[152]) This interesting possibility for CH_3F can be neither confirmed nor refuted by the fluorescence measurements, and the relaxation time predicted by SSH theory for the ν_6 mode is 19 300 ns, an order of magnitude larger than observed. Closer agreement is shown for CH_3Cl, where better intermolecular force constants are available, and a V-T relaxation time of 495 ns is predicted. Other simple polyatomic molecules where laser fluorescence studies have been made and are in accord with the model of rapid inter-mode V-V transfer, followed by V-T relaxation of the total vibrational energy through a single mode, are SF_6,[264] O_3,[265] CH_2F_2,[266] and CO_2.[337]

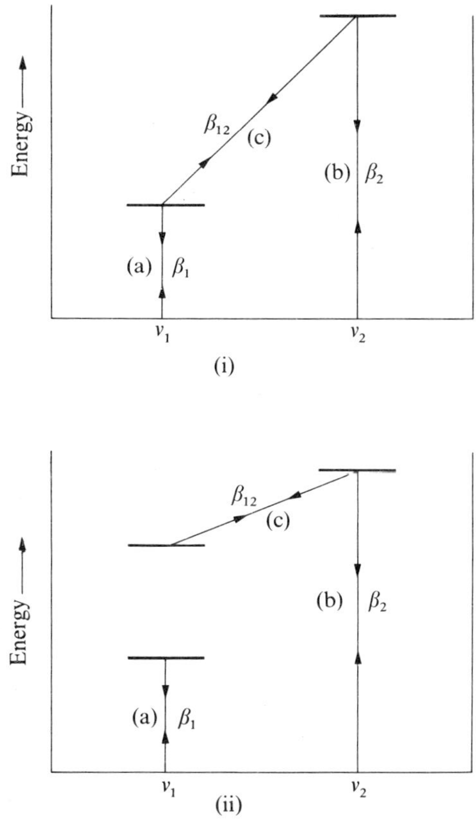

FIG. 4.5. Energy-level diagram for a molecule with two active vibrational modes, showing double relaxation process

4.2.2. *Multiple relaxation process*

A few polyatomic molecules have a large energy gap between the lowest and second lowest vibrational modes. The vibrational transitions for such a molecule, having two active vibrational modes ν_1 and ν_2, are shown in Fig. 4.5(i), which may be contrasted with Fig. 4.1. The same three transitions, (a), (b), and (c) are possible, but similar reasoning now gives the condition $\beta_2 \gg \beta_{12} > \beta_1$. Process (b) will again be too slow to play any role, and the upper mode will again relax by a series mechanism via the complex process (c), followed by (a). But (c) is now slower than (a) and becomes the rate-determining step, so that the vibrational energy of the upper mode is transferred to translation with an effective relaxation time β_{12}. The vibrational energy of the lowest mode relaxes independently by process (a) with the shorter relaxation time β_1. For the crude model shown in Fig. 4.5(i) the criterion for a multiple relaxation process of this kind would be the simple condition $\nu_2 > 2\nu_1$. Actually the fastest complex process for relaxation of the upper mode will be a transition to the *first harmonic* ($2\nu_1$) of the lower mode, as shown in Fig. 4.5(ii):

$$A^*(\nu_2) + A(0) \rightarrow A^*(2\nu_1) + A(0) + \Delta E,$$

followed by very rapid resonant single-quantum (intermolecular) transfer:

$$A^*(2\nu_1) + A(0) \rightarrow A^*(\nu_1) + A^*(\nu_1).$$

The first step is a two-quantum transition which is intrinsically slower than a one-quantum transition (cf. § 3.1.3 (c)), but will still be faster than (a) for *small* values of $\Delta E = \nu_2 - 2\nu_1$, so that the criterion becomes $\nu_2 \gg 2\nu_1$.

The most widely investigated example of this behaviour is SO_2, whose energy levels are shown in Fig. 4.6. Ultrasonic speed measurements at 300 K[183] show a clear *double*-dispersion curve, Fig. 4.7, characterized by two relaxation times: $\beta_1 = 50$ ns, corresponding to the vibrational specific heat $C_1 = 1{\cdot}19$ cal K^{-1} mol^{-1} of the lowest mode, ν_2: $\beta_2 = 1250$ ns corresponding to a specific heat $C_2 = 0{\cdot}34$ cal K^{-1} mol^{-1} which is the sum of the specific heat contributions of the two upper modes ν_1 and ν_3. The inference follows that there is rapid equilibration between the two upper modes ν_1 and ν_3, and that their combined vibrational energy relaxes by the rate-determining complex process

$$SO_2^*(\nu_1) + SO_2(0) \rightarrow SO_2^*(2\nu_2) + SO_2(0) + 110\ \text{cm}^{-1},$$

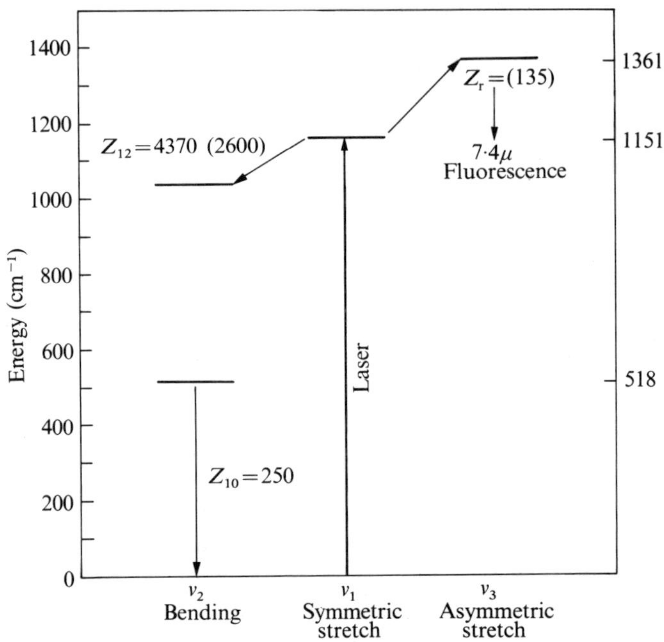

FIG. 4.6. Collision numbers for vibrational transitions in SO_2. Numbers in parentheses from fluorescence;[270] numbers without parentheses from ultrasonic dispersion[183]

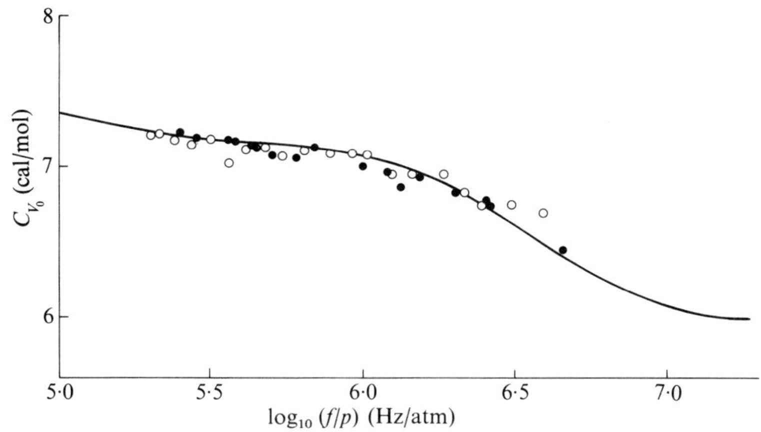

FIG. 4.7. Acoustic dispersion curve for SO_2. Open circles are observed values from ref. 183; filled circles are observed values from ref. 159 (recalculated ref. 183). The solid curve is theoretical, calculated for the double relaxation process: $\beta_1 = 50$ ns, $C_1' = 1{\cdot}19$ cal; $\beta_2 = 1250$ ns, $C_2' = 0{\cdot}34$ cal

followed by the rapid resonant transition:

$$SO_2{}^*(2\nu_2) + SO_2(0) \rightarrow SO_2(\nu_2) + SO_2(\nu_2).$$

The collision number $Z_{12} = 4370$ is calculated from β_2 on this assumption. The lower (bending) mode ν_2 relaxes by simple V-T transfer, giving $Z_{10} = 250$. This series mechanism (in contrast to a 'parallel' mechanism, where ν_2 and ν_1 both relax independently to translation) is in accord with further extensive ultrasonic absorption measurements by Shields and Anderson[267] on mixtures of SO_2 + Ar.

Theoretical SSH calculations,[183] based on the 12-6-3 intermolecular potential of Monchick and Mason,[268] predict this mechanism, but the calculated relaxation times $\beta_1 = 201$ ns and $\beta_2 = 2805$ ns are much too high. (Earlier calculations by Dickens and Linnett based on a Krieger potential predict still higher relaxation times.[269]) Laser fluorescence experiments on SO_2 have been made by Siebert and Flynn[270] using a similar technique to that described for CH_3Cl and CH_3F in § 4.2.2. They pumped the ν_1 mode by a *Q*-switched CO_2 laser pulse, and observed the risetime of fluorescence for the ν_3 mode. This showed rapid energy transfer, corresponding to a collision number $Z_r = 135$, ascribed to the rapid single-quantum transition:

$$SO_2{}^*(\nu_1) + SO_2(0) \rightarrow SO_2(0) + SO_2(\nu_3) - 210 \text{ cm}^{-1}.$$

This *endothermic* process leads to rapid kinetic *cooling* at the centre of the laser beam, which can be observed as 'thermal lensing',[271] similar to that described in § 2.5 for rapid kinetic *heating*. Equilibrium between ν_1 and ν_3 is thus rapidly established, and the subsequent fluorescence decay rate of ν_3 corresponds to relaxation of the energy of the upper vibrational manifold, $(\nu_1 + \nu_3)$, mainly by the process

$$SO_2^* (\nu_1) + SO_2 (0) \rightarrow SO_2 (0) + SO_2 (2\nu_2) + 110 \text{ cm}^{-1}.$$

This corresponds to a collision number $Z_{12} = 2600$ which is lower than the ultrasonic value. The fluorescent decay of ν_2 could not be observed.

Double ultrasonic dispersion curves have been observed for five other molecules, and the results are summarized in Table 4.1. For SO_2 the difference

TABLE 4.1.

Experimental collision numbers for intramolecular vibrational energy transfer at 294 K (ref. 183)

Substance	$\nu(1)$ (cm^{-1})	$\nu(2)$ (cm^{-1})	i	$\Delta\nu$ (cm^{-1})	Z_{12}	Z_{10}
CF_3CN (ref. 183)	187	464	2	−90	60	10
C_2H_6 (ref. 174)	290	820	3	50	90	10
C_2N_2 (ref. 183)	226	506	2	−54	220	18
C_3O_2 (ref. 33)	63	550	9	−17	460	3
CH_2Cl_2 (ref. 175)	283	704	2	−140	460	30
SO_2 (ref. 183)	519	1151	2	−110	4370	250

Z_{12} is the collision number for V-V transfer between one quantum of the second lowest mode, $\nu(2)$ and i quanta of the lowest mode, $\nu(1)$.
Z_{10} is the collision number for V-T transfer from $\nu(1)$.

between β_1 and β_2 is large enough for the two relaxation processes to be treated independently, as above. For the other molecules the two processes are *not* independent, and the more sophisticated treatment of the dispersion curves, described at the end of § 2.2.1 has been used. The energy levels are shown in Fig. 4.8, together with those of CH_2Br_2 for which a double relaxation process was predicted from SSH calculations by Dickens and Schofield,[272] though ultrasonic measurements have been indecisive.[273,274] For C_2N_2 the energy gap between the third and second modes is even larger than that between the second and first. SSH calculations by Dickens[275] predict *three* relaxation times, essentially one for each mode, but ultrasonic dispersion measurements show no trace of the third (and longest) relaxation time predicted for the third mode. It seems that, in general, SSH theory overestimates the relaxation times for V-V transfer between upper modes, and may predict double relaxation which is not found experimentally (e.g. $CHCl_3$ and CCl_4).[27] It is interesting to note that the lowest mode of C_2H_6 is the torsional oscillation ν_4, due to restricted internal rotation of the CH_3 groups, and that this seems to relax in just the same way as stretching and bending modes. The general trend of Z_{12} in Table 4.1 shows the positive dependence on ν, i, and $\Delta\nu$ predicted by SSH theory. The anamalously low value of $Z_{12} = 90$ for C_2H_6 ($\nu(2) = 820$ cm^{-1},

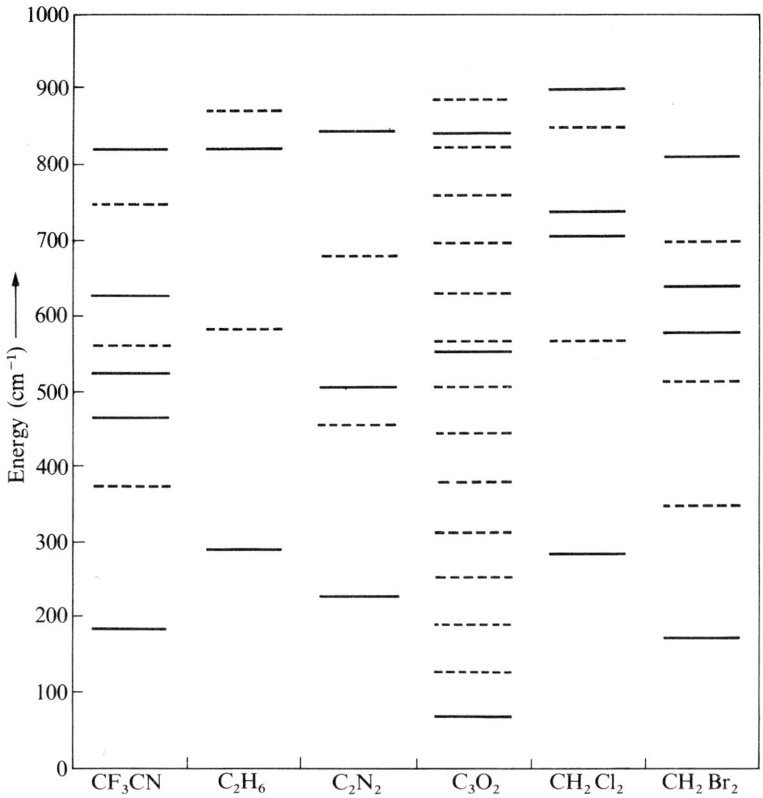

FIG. 4.8. Vibrational energy levels for molecules showing multiple relaxation (for SO_2 see Fig. 4.6). Solid lines, fundamental modes; dashed lines, harmonics of lowest mode

i=3, $\Delta\nu = +50\ cm^{-1}$), as contrasted with $Z_{12} = 220$ for C_2N_2 ($\nu(2) = 506\ cm^{-1}$, i=2, $\Delta\nu = -54\ cm^{-1}$), may be due to the hydrogen-effect (cf. § 3.4.2). For C_3O_2 the very low value of $\Delta\nu = -17\ cm^{-1}$ is counterbalanced by i=9, leading to the high value $Z_{12} = 460$.

4.3. Ultrasonic dispersion in polyatomic mixtures

The nature of the overall relaxation process for a binary mixture of a relaxing gas, A, with a non-relaxing gas, B, was discussed in § 3.5.1. Two collision processes are involved:

$$\text{(1)}\quad A^* + A \rightarrow A + A \qquad \text{V-T}$$
$$\text{(2)}\quad A^* + B \rightarrow A + B \qquad \text{V-T,}$$

and the overall relaxation time is governed by eqn (3.15), giving a linear plot of reciprocal relaxation time, $1/\beta_A$, against x, the mole fraction of B in the mixture (Fig. 3.15).

If both A and B are polyatomic relaxing gases, there will also be two collision processes, corresponding to (1) and (2), for V-T energy transfer from B* in homomolecular and heteromolecular collisions. In addition there can be a V-V transfer between A* and B*, making five transfer processes in all:

(1) $A^* + A \rightarrow A + A$ V-T
(2) $A^* + B \rightarrow A + B$ V-T
(3) $A^* + B \rightarrow A + B^*$ V-(V±T)
(4) $B^* + B \rightarrow B + B$ V-T
(5) $B^* + A \rightarrow B + A$ V-T.

If process (3), V-V transfer, does not occur, the mixture will show a double relaxation process, similar to that given for SO_2 in Fig. 4.7, characterized by two relaxation times β_A and β_B, which will be related to molar composition by equations of type (3.15), each giving a linear plot of $1/\beta$ against composition. This is a *parallel* mechanism, and is represented on the energy-level diagram in Fig. 4.9. If V-V transfer does occur, the picture is completely altered. Supposing,

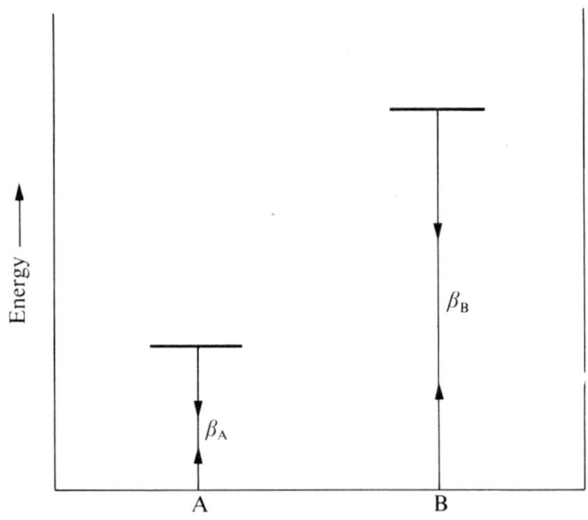

FIG. 4.9. Energy-level diagram showing transitions for a mixture of two relaxing gases without V-V transfer (parallel mechanism)

for convenience, that pure A relaxes slowly, and pure B rapidly, so that processes (4) and (5) are both much faster than processes (1) and (2), there are now two alternative possibilities for the overall relaxation process.

When V-V transfer is faster than all the other processes, (3) will maintain the vibrational energy of the whole system ($A^* + B^*$) in continuous equilibrium, and the total vibrational heat content of the mixture will relax with a single relaxation time *via* the faster of processes (4) or (5). If (4) plays the predominant role, this will give rise to a near *quadratic* dependence of $1/\beta$ on mole fraction of B, since the rate of process (4) is proportional to x^2, and eqn (3.15) no longer applies. The rapid *near-resonant* collision process (3), involving transfer of vibrational energy from a mode of frequency ν_i of molecule A to a mode of frequency ν_j of molecule B, plays exactly the same role as the *complex* collision process, involving transfer of vibrational energy from mode i to mode j of a single molecular species. This is a *series* mechanism, similar to that shown in Fig. 4.1 for a pure polyatomic gas showing a single relaxation time.

Alternatively, when process (3) is slower than (4) or (5), but faster than (1) or (2), A will again relax by route (3) followed by (4) or (5), but now (3) will be rate determining. This will give a *linear* variation of $1/\beta_A$ with x. B will relax independently, and more rapidly, *via* (4) and (5), with linear dependence of $1/\beta_B$ on x. There will be a double relaxation process with two relaxation times, β_A involving only the vibrational heat capacity of A, and β_B only that of B, both showing linear concentration dependence. This mechanism is analogous to the relaxation behaviour discussed in § 4.2.2 for pure polyatomic gases which show double dispersion because V-V transfer between modes is *slower* than V-T transfer from the lowest mode. This is again a series mechanism similar to that shown in Fig. 4.5 for pure polyatomic gases giving double dispersion.

The nature of the overall relaxation process (whether single or double) for mixtures, and the concentration dependence of the relaxation times, are thus determined by the relative rates of processes (1) to (5). Observations of ultrasonic dispersion in the two pure gases and in a series of mixtures extending over the whole concentration range enable a diagnosis to be made of which type of mechanism is operating. The rates of processes (1) and (4) are obtained from measurements on the pure components, and the rates of processes (3) and (5) which give the best fit to the experimental observations may be estimated by trial and error. Results obtained in this way by Lambert, Parks-Smith, and Stretton[182] for a series of mixtures are shown in Table 4.2. Mixtures were chosen for which near-resonant vibration frequencies in the two components would be expected to give rise to rapid V-V transfer (small value of Z_{AB}). It will be seen that, where Z_{AB} lies *below* the collision numbers for V-T relaxation of the pure components, Z_{AA} and Z_{BB}, single dispersion is observed; where Z_{AB} lies *between* Z_{AA} and Z_{BB}, double dispersion is observed.

The actual experimental data for one singly dispersing mixture, $SF_6 + CHClF_2$, are shown in Fig. 4.10. The lowest vibrational modes of the two molecules lie close enough for rapid V-V transfer, and single dispersion is observed, with near-quadratic concentration dependence of $1/\beta$. This indicates that the homomolecular relaxation of $CHClF_2$ (B), process (4), is the rate-controlling step, and is faster than the heteromolecular relaxation process (5). The curve calculated for concentration dependence of $1/\beta$ was obtained by setting up the detailed energy and temperature relaxation equations developed by Tanczos,[27] and solving over the whole concentration range by computer. The value of Z_{AB} in Table 4.2 was estimated to give the best fit, and may be taken as 50±15. Relaxation behaviour

TABLE 4.2.

Experimental collision numbers for intermolecular vibrational energy transfer at 300 K (ref. 182)

A	B	ν_A (cm^{-1})	ν_B (cm^{-1})	i	$\Delta\nu$ (cm^{-1})	Z_{AB}	Z_{AA}	Z_{BB}
			Singly dispersing mixtures					
SF_6	$CHClF_2$	344	369	1	25	50	1005	122
C_2H_4	C_2H_6	810	821·5	1	11·5	40	970	74
			Doubly dispersing mixtures					
CCl_2F_2	CH_3OCH_3	260	250	1	10	5	73	<3
CH_3Cl	CH_3OCH_3	732	250	3	18	70	421	<3
SF_6	CH_3OCH_3	344	164	2	16	80	1005	<3
CHF_3	C_2F_4	507	507	1	0	50	1500	5·5
SF_6	C_2F_4	344	190	2	36	70	1005	5·5
CF^4	C_2F_4	435	220	2	5	110	2330	5·5

Z_{AB} is the collision number for vibration-vibration transfer between one quantum of mode of frequency ν_A of molecule A and i quanta of mode of frequency ν_B of molecule B. (Frequencies are for 0-1 vibrational excitation.)

of the same type has also been reported by Rao and Srinivasaschari[277] for mixtures of CO_2 with C_2H_4O. Here rapid V-V exchange is assumed to occur between the 667·3 cm^{-1} mode of CO_2 and the 704 cm^{-1} mode of C_2H_4O, and all mixtures show single dispersion with near-quadratic dependence of $1/\beta$. Detailed calculations have not been made.

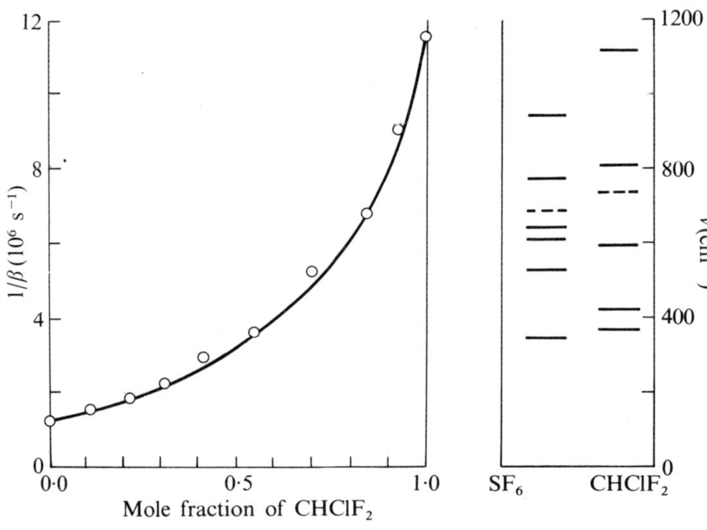

FIG. 4.10. Reciprocal relaxation times and energy-level diagram for SF_6 + $CHClF_2$ mixtures (ref. 182). Open circles, observed values; solid curve calculated from theory

The other mixture in Table 4.2 showing single dispersion is $C_2H_4 + C_2H_6$, investigated experimentally by Valley and Legvold.[276] This case is complicated by the double dispersion shown by pure C_2H_6 (see § 4.2.2), and the 'single dispersion' involves only the upper vibrational manifold of C_2H_6. The data are shown in Fig. 4.11. There is near-resonance between the second lowest mode

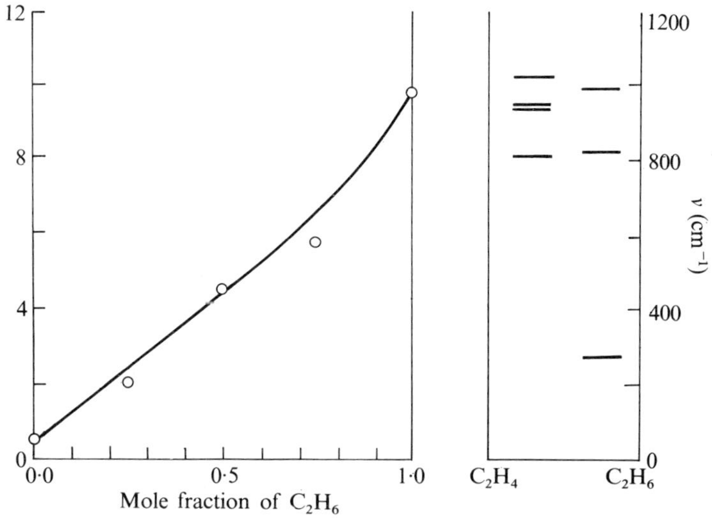

FIG. 4.11. Reciprocal relaxation times and energy-level diagram for C_2H_4 + C_2H_6 mixtures (ref. 182). Open circles, observed values recalculated from ref. 276; solid curve calculated from theory

of C_2H_6 (821 cm^{-1}) and the lowest mode of C_2H_4 (810 cm^{-1}), leading to a single dispersion involving the whole vibrational energy of C_2H_4 and the upper manifold of C_2H_6. Z_{AB} is the collision number for V-V transfer between the two modes, Z_{BB} for V-T relaxation of the 821 cm^{-1} mode of C_2H_6, and Z_{AA} for the 810 cm^{-1} mode of C_2H_4. The efficiencies of homomolecular and heteromolecular V-T relaxation of C_2H_6 are roughly the same, and an almost linear concentration dependence of $1/\beta$ is observed. The torsional (290 cm^{-1}) mode of C_2H_6 relaxes independently in both pure gas and mixtures, so that the mixtures actually show an *overall* double dispersion.

This behaviour is in striking contrast to that of mixtures of the doubly dispersing gas SO_2 with CH_2F_2, which were later investigated by Lambert *et al.*,[183] and show a clear single relaxation process for the total vibrational energy of both molecules (Fig. 4.12), (contrast Fig. 4.7 for pure SO_2). The energy levels for the system are shown in Fig. 4.13. Pure CH_2F_2 shows a single relaxation process, owing to near-resonance between ν_3 and $2\nu_4$ leading to rapid inter-mode V-V transfer, which is faster than the V-T relaxation of ν_4.[155,183] The suppression of the separate relaxation time for ν_1 of SO_2 is due to rapid near-

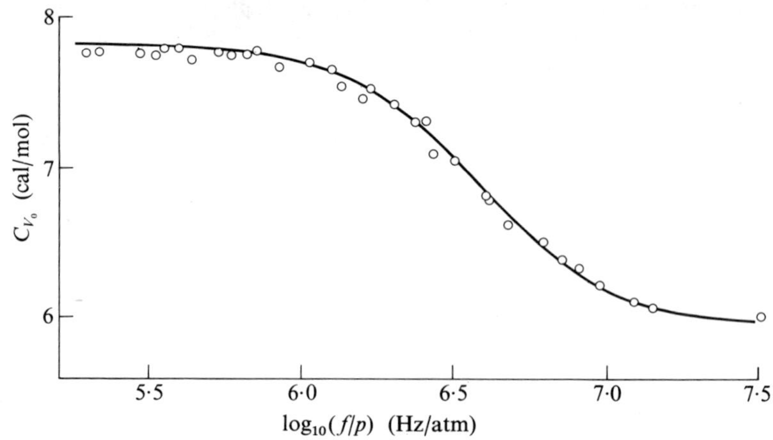

FIG. 4.12. Acoustic dispersion curve for mixture of SO_2 with 40 mol per cent CH_2F_2. Open circles, observed values (ref. 183). The solid line is the theoretical curve calculated for single relaxation process: $\beta = 49{\cdot}1$ ns

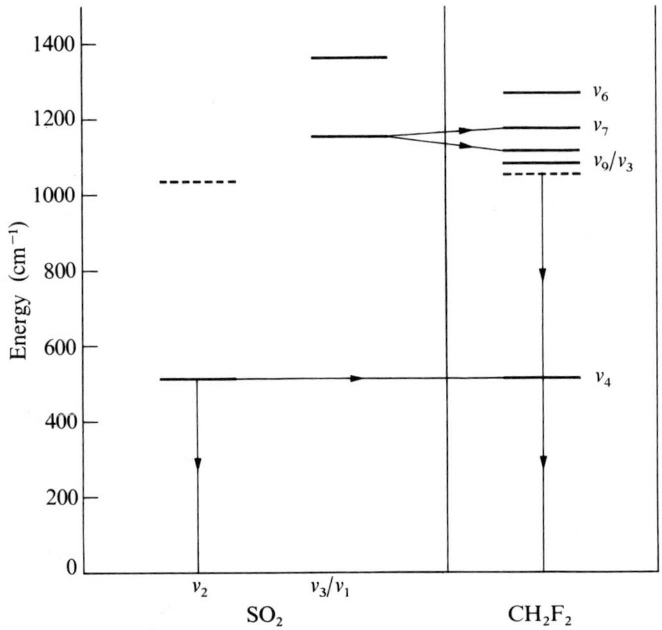

FIG. 4.13. Vibrational energy levels and transitions for SO_2 + CH_2F_2. Solid lines are fundamental modes; dashed lines are harmonics

resonant V-V transfers:

$$SO_2(\nu_1) + CH_2F_2(0) \rightarrow SO_2(0) + CH_2F_2(\nu_3) + 35\ \text{cm}^{-1}$$

and

$$SO_2(\nu_1) + CH_2F_2(0) \rightarrow SO_2(0) + CH_2F_2(\nu_7) - 31\ \text{cm}^{-1},$$

followed by rapid intramolecular relaxation of the ν_3 and ν_9 modes of CH_2F_2. There is also near-resonance between the two lowest modes:

$$SO_2(\nu_2) + CH_2F_2(0) \rightarrow SO_2(0) + CH_2F_2(\nu_4) - 10\ cm^{-1}.$$

The V-T relaxation of ν_4 of the hydrogen-containing CH_2F_2 ($\beta = 27$ ns) is much faster than that of the slightly lower ν_2 of SO_2 ($\beta_1 = 50$ ns), so that the whole vibrational energy of both molecules is maintained in continuous dynamic equilibrium and relaxes in a single combined V-T process, mainly *via* ν_4 of CH_2F_2. This furnishes an interesting example of a molecule, which itself has a multiplicity of vibrational modes and shows rapid intramolecular V-V transfer, acting as an energy transfer *catalyst* in promoting faster V-V transfer between the more widely spaced modes of a doubly relaxing molecule. The system is too complicated for meaningful estimates of Z_{AB} to be made.

The mixtures in the second section of Table 4.2 (which were investigated earlier[278] when erroneous conclusions were drawn) all show genuine double dispersion. The data for one mixture, $SF_6 + C_2F_4$, are shown in Fig. 4.14.

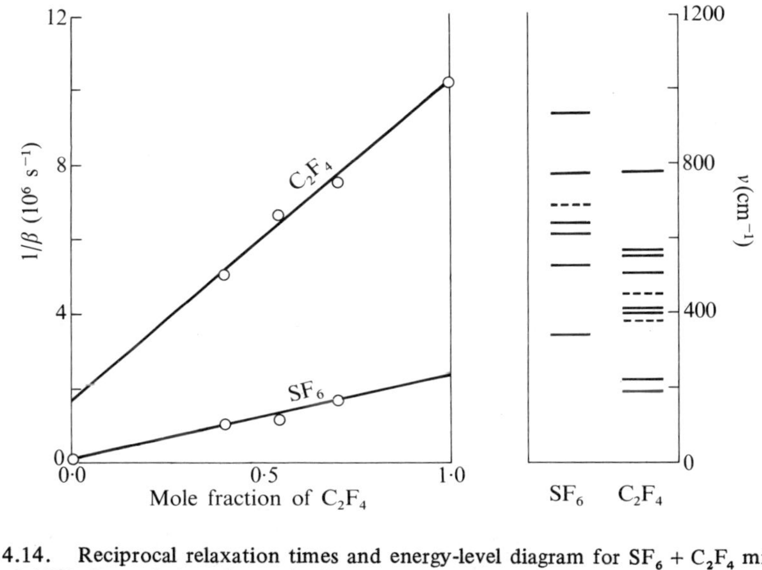

FIG. 4.14. Reciprocal relaxation times and energy-level diagram for $SF_6 + C_2F_4$ mixtures (ref. 182). Open circles, observed values

There is near-resonance between the lowest (344 cm^{-1}) mode of SF_6 and the first harmonic of the lowest (190 cm^{-1}) mode of C_2F_4. C_2F_4 shows very efficient homomolecular V-T transfer, and the estimated V-V transfer rate ($Z_{AB} = 70$) falls between this and the slower V-T transfer rate of SF_6 ($Z_{AA} = 1005$). Double dispersion is observed, and the predicted linear variation with concentration of the two relaxation times. The remaining mixtures in this section, all of which involve B components whose homomolecular relaxation is very rapid, behave similarly.

Detailed SSH calculations are not feasible for most of the large molecules listed in Table 4.2, but the general trend of Z_{AB} values is again roughly in accord with theoretical prediction. The largest value of $\Delta\nu$ is 36 cm^{-1}, so that all the transfers may be regarded as resonant and follow eqn (4.2). The collision efficiency should be independent of $\Delta\nu$, and decrease with rising frequency of the exchanging modes and with increase in i (multi-quantum transfer). The lowest value of $Z_{AB} = 5$ is for a one-quantum transition between CCl_2F_2 and CH_3OCH_3 ($\nu = 250$ cm^{-1}): the other one-quantum transfers involve higher frequency modes and show values of $Z_{AB} \sim 50$. The mixture $C_2H_4 + C_2H_6$ with the highest value of $\nu = 821{\cdot}5$ cm^{-1}, gives an unexpectedly high efficiency ($Z_{AB} = 40$), and this may again be attributed to the hydrogen-effect (cf. §§ 3.4.2 and 4.2.2). The efficiencies of two- and three-quantum transfers (Z_{AB} ranging from 70 to 110) are all lower than the one-quantum transfers (Z_{AB} ranging from 5 to 50). There is no discernible dependence on $\Delta\nu$, and the exactly resonant, one-quantum transfer, $CHF_3 + C_2F_4$ ($\nu = 507$ cm^{-1}), shows $Z_{AB} = 50$.

For diatomic and simpler polyatomic molecules the necessary data for full *a priori* SSH calculations are often available. Calculations made by Lambert *et al.*[279] for some O_2 mixtures are shown in Table 4.3. The results fall into the same kind of pattern discussed above, and show the striking 'catalytic' effects of H_2O and hydrocarbons. Acoustic measurements have only been made at additive concentrations too low to enable accurate collision numbers to be estimated, but they appear to be in general accord with the theoretical predictions. A full discussion is given in ref. 279.

TABLE 4.3.

Calculated collision numbers for intermolecular vibrational energy transfer in oxygen mixtures at 300 K (ref. 279)

Additive (B)	ν_B (cm^{-1})	i	$\Delta\nu$ (cm^{-1})	Z_{AB}	Z_{BB}
CH_4	1534	1	20	170	11600
CD_4	1092	1	462	16000	4960
C_2H_4	1444	1	110	490	4050
C_2H_2	729	2	96	1800	490
H_2O	1596	1	42	80	~1
D_2O	1178	1	376	1160	~1
HDO	1402	1	152	140	~1

Z_{AB} is the collision number for vibration-vibration transfer between one quantum of the fundamental mode of O_2, $\nu_A = 1554$ cm^{-1}, and i quanta of mode of frequency ν_B of molecule B. For pure O_2, $Z_{AA} = 8{\cdot}31 \times 10^7$. Collision parameters for heteromolecular collisions calculated using combining rules of Hirschfelder *et al.*[1]

4.4. Relaxation of infrared active modes (role of transition dipole)

4.4.1. Non-polar molecules. CO ($v \geqslant 1$) and CO_2 ($00^0 1$)

CO is one of the few almost non-polar diatomic gases which is infrared active, so that its vibrational relaxation may be followed by spectroscopic techniques. V-T relaxation, discussed in § 3.3.1, is very slow ($Z_{10} > 10^9$ at 300 K) and the much faster V-V relaxation is easy to study, leading to a wide range of available data. Vibrationally excited CO ($4 \leqslant v \leqslant 13$) is produced by the reaction O + CS → CO* + S, and Hancock and Smith[66] have made extensive studies of the decay of infrared chemiluminescence due to subsequent processes,

$$\mathrm{CO}\ (v) + \mathrm{AB}\ (v{=}0) \rightarrow \mathrm{CO}\ (v{-}1) + \mathrm{AB}\ (v{=}1) \pm \Delta E,$$

where AB may be another CO molecule or an additive. Owing to anharmonicity the values of ΔE change substantially with change in v, so that the role played by variation in both parameters may be investigated for each single collision pair. The energy-level diagram for the systems investigated is given in Fig. 4.15, and it will be seen that for two systems, CO + COS and CO + NO, ΔE actually changes sign as v increases. SSH theory predicts that for both resonant and non-resonant transfer (eqns (4.2) and (3.2) respectively), the probability of a transition between v and $(v{-}1)$ increases proportionately to v, i.e. $P_{v,v-1} = vP_{1,0}$ (see § 3.1.3(c)). The collision number Z_{AB} ($= 1/P_{\mathrm{AB}}$) will decrease in the

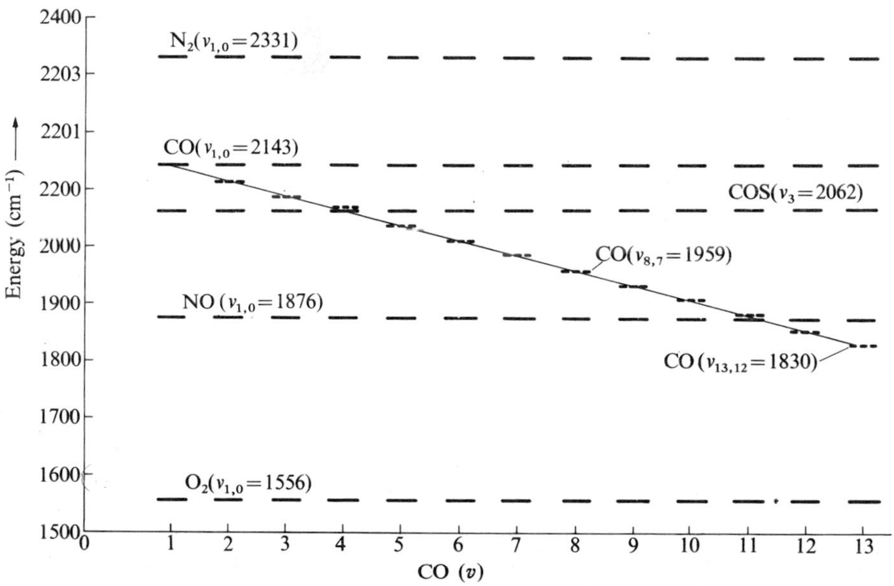

FIG. 4.15. Energy-level diagram for single-quantum transfer from CO(v) to additive, AB(v=0)

same proportion, and the relevant parameter for comparing the different transitions is the *reduced collision number*, $vZ_{AB}(=v/P_{AB})$.

Fig. 4.16 shows a plot of log vZ_{AB} against ΔE: the collision numbers have all been calculated for transition in the *exothermic* direction, using the relation $P_{01} = P_{10} \exp(-\Delta E/RT)$ where necessary. Laser fluorescence data for one-quantum transfer between CO (v=1) and N_2 and NO are included. The plot for each pair shows qualitative agreement with the major features predicted by theory. For values of $\Delta E \leqslant 50$ cm^{-1}, where eqn (4.2) is applicable (CO + CO; CO + NO; CO + COS), the value of vZ_{AB} is constant, and independent of ΔE. For values of $\Delta E \geqslant 200$ cm^{-1}, where eqn (3.2) becomes applicable, the predicted linear dependence of log vZ_{AB} on ΔE is found. The slopes of the plots for the five collision partners, O_2, N_2, CO, NO, COS are the same and equal to that of the Lambert-Salter plot for V-T transfer in non-hydrogen-containing molecules (Fig. 3.12), corresponding to a value $C = 17 \times 10^{-3}$ cm of the gradient in eqn (3.12). The same slope was shown earlier by Callear[298] to fit a similar plot for V-V transfer between NO $X^2\Pi$ (v=1) and CO $X^1\Sigma$, N_2 $X^1\Sigma$ or NO $A^2\Sigma$. For values of ΔE lying between 50 and 200 cm^{-1}, where neither of the analytical solutions of the SSH equations is applicable, all the mixtures show a similar, rather steeper, gradient.

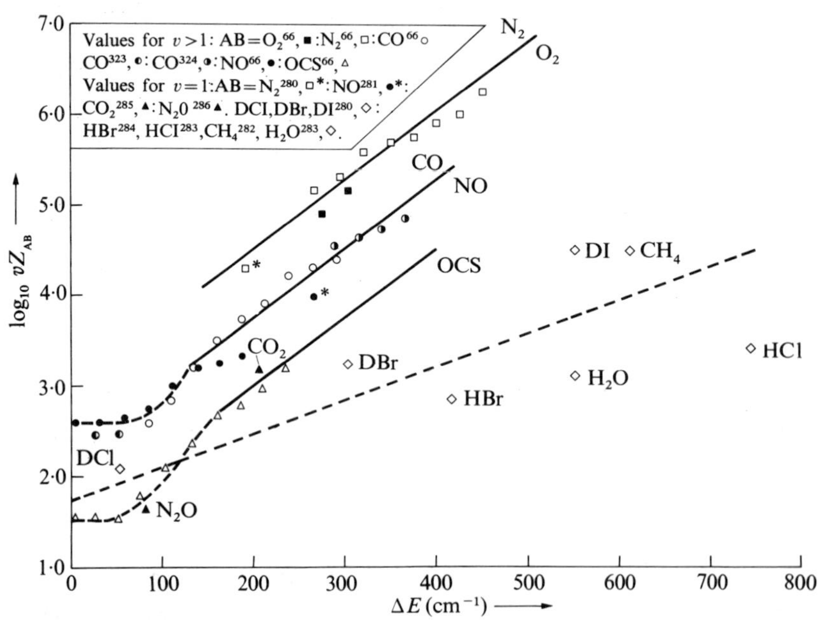

FIG. 4.16. Reduced collision numbers for V-V transfer at 300 K $CO(v) + AB(0) \rightarrow CO(v-1) + AB(v=1) \pm \Delta E$.
All values calculated for transition in the exothermic direction. Solid line is the slope of Lambert-Salter plot II (fig. 3.12). Dashed line is the slope of Lambert-Salter plot I (Fig. 3.11)

The absolute magnitudes of the reduced collision numbers fall into three distinct groups, differing by almost an order of magnitude: collisions with N_2 and O_2 are the least efficient, with NO and CO the next, and with OCS the most efficient. This progression does not correlate in any obvious way with the major factors in SSH theory; vibrational frequency of the exchanging modes, collision mass, or intermolecular forces. If it is assumed that for OCS the ν_3 mode is that involved in the transfer, the variation in efficiency between the three groups runs parallel with the variation in the *infrared activities* of the collision partners: COS $(\nu_3) >$ CO, NO, $> N_2$, O_2 (infrared inactive). This is in accord with the suggestion of Mahan,[292] that V-V transfer between two infrared active modes may be brought about by a factor which does not operate in V-T transfer, long-range dipole-dipole coupling between the *transition dipoles* of the two molecules. This leads to a transition probability depending on the product of the squares of the transition dipole moments and having a negative temperature dependence. The theory has been developed by Sharma,[293] and calculations for the CO+CO and CO+NO systems show fair agreement with experiment when $\Delta E \leqslant 100$ cm^{-1}, but predict too steep a slope at higher values of ΔE.[66] This failure was attributed by Jeffers and Kelley[325] to the neglect of the short-range repulsive interactions, which become more important as the transitions become less resonant. Their theory, which includes both long-range and short-range forces, gives better agreement for CO+CO at 300 K, but is less successful at 100 K.[326] A more sophisticated quantum-mechanical treatment by Dillon and Stephenson,[327] who show that for higher vibrational states with large transition moments, multi-quantum processes, both rotational and vibrational, play a significant role, seems more successful overall.

For the mixtures with N_2 and O_2, both of which are infrared inactive, this effect cannot operate, and simple SSH calculations by Stretton and Parks-Smith[297] for the transition

$$N_2(v{=}1) + CO(v{=}0) \rightarrow N_2(v{=}0) + CO(v{=}1) + 188\ \text{cm}^{-1},$$

give a value of $Z_{AB} = 2{\cdot}2 \times 10^4$ at 300 K in good agreement with the experimental (laser fluorescence) value $Z_{AB} = 2{\cdot}0 \times 10^4$.[280] The same applies for the more strongly exothermic transition

$$CO(v{=}1) + CH_4(0) \rightarrow CO(v{=}0) + CH_4(\nu_2) + 617\ \text{cm}^{-1}.$$

The ν_2 mode of CH_4 is not infrared active, and Millikan's experimental value, $Z_{AB} = 3{\cdot}3 \times 10^4$, is in fair agreement with the SSH calculated value, $Z_{AB} = 2{\cdot}1 \times 10^4$.[294] In contrast, SSH calculations for mixtures with the infrared-active NO give $Z_{AB} = 3{\cdot}68 \times 10^4$,[29] which is significantly higher than the experimental value, $Z_{AB} = 1{\cdot}0 \times 10^4$, in accord with the view that an additional factor is operating here.

Confirmation of the importance of long-range forces between transition dipoles is given by the relaxation behaviour of mixtures with CO_2 and with N_2). The relevant vibrational states of the two molecules are:

	CO_2				N_2O		
ν_2	(010)	667 cm^{-1}	i.r.active	ν_2	(010)	589	i.r.active
ν_1	(100)	1388	i.r.inactive	ν_1	(100)	1285	i.r.active
$2\nu_2$	(020)	1285	i.r.inactive				
ν_3	(001)	2349	i.r.active	ν_3	(001)	2223	i.r.active

V-V transfer from CO (ν_{10} = 2143) involves the near-resonant ν_3 modes of both molecules:

$$CO\ (v=1) + CO_2(000) \rightarrow CO\ (v=0) + CO_2(001) - 206\ cm^{-1}$$

$$CO\ (v=1) + N_2O(000) \rightarrow CO\ (v=0) + N_2O(001) - 80\ cm^{-1}.$$

Both ν_3 modes are strongly infrared active and the experimental transfer efficiencies,[285,286] which are plotted in Fig. 4.16, are close to those for OCS (ν_3). V-V transfer from the related infrared-active species CS (ν_{10} = 1272 cm^{-1}) to CO_2 and N_2O has been investigated by Morley and Smith.[310] Here there are closely resonant transfers to the two 1285 cm^{-1} levels:

$$CS\ (v=1) + CO_2(000) \rightarrow CS\ (v=0) + CO_2(020) - 13\ cm^{-1}$$

$$CS\ (v=1) + N_2O(000) \rightarrow CS\ (v=0) + N_2O(100) - 13\ cm^{-1},$$

but the N_2O transition (000) to (100) is strongly infrared active, whilst the CO_2 (000) to (020) (also to the Fermi resonant (100) level) is not. Transfer to N_2O, Z_{AB}=26, is two orders of magnitude faster than transfer to CO_2, Z_{AB} = 2700, in spite of identical values of ΔE.

Laser fluorescence data for single-quantum transitions between CO(v=1) and hydrogen and deuterium halides together with H_2O and CH_4 are also shown on Fig. 4.16. These gases are all strongly polar (Table 3.3) and infrared active, and show much more efficient transfer at high values of ΔE. The deuterides are consistently less efficient than the hydrides, and the Z_{AB} values for both classes scatter on either side of a line drawn with the Lambert-Salter slope for V-T transfer in polyatomic hydrides (Fig. 3.11). The near-resonant CO+DCl transition shows an efficiency comparable with CO+COS. The hydrides and deuterides all possess very low moments of inertia, so that V-R transfer is likely to play an important additional role, and theoretical calculations would be difficult. The general resemblance of the gradients in Fig. 4.16 to the Lambert-Salter plots for V-T transfer in polyatomic molecules is unlikely to be simply fortuitous. Procaccia and Levine[299] have extended the information theoretic approach discussed in § 3.6 to V-V transfer, and have shown that a similar analysis, leading to a similar expression for dependence on ΔE, applies here too.

Some studies of the temperature-dependence of V-V transfer efficiency for CO mixtures are shown on a Landau-Teller type plot in Fig. 4.17, and are in general accord with theoretical prediction. The exothermic N_2+CO transfer (ΔE = 188 cm^{-1}), where only short-range repulsive forces are involved, shows a normal positive $T^{-\frac{1}{3}}$ dependence, and excellent agreement with SSH calculations.

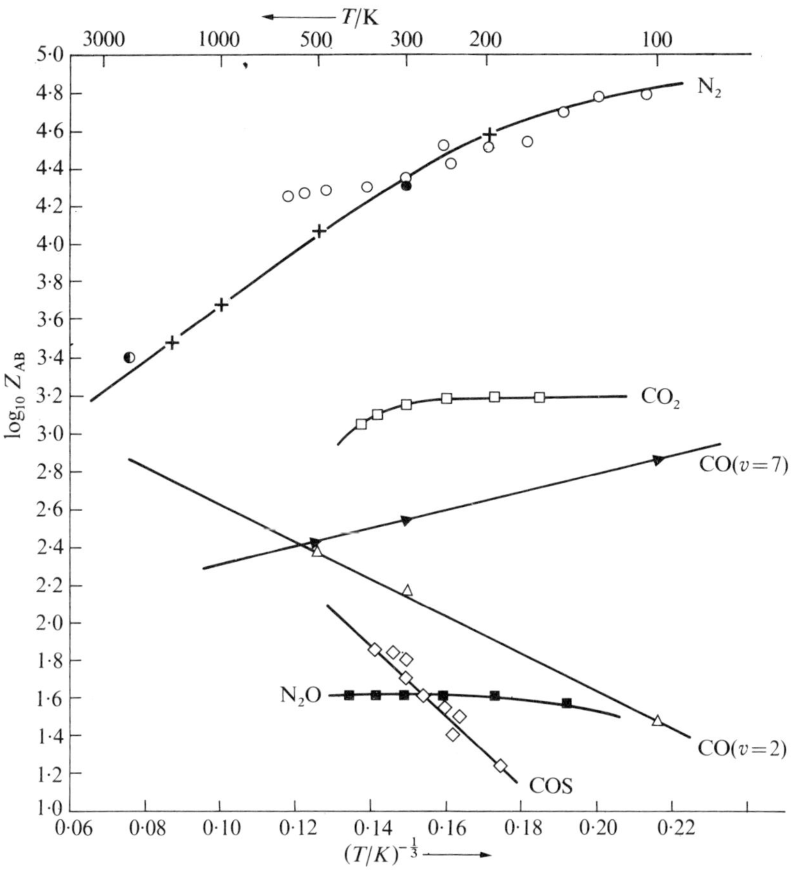

$N_2(v=1)+CO(v=0) \longrightarrow N_2(v=0)+CO(v=1)+188 cm^{-1}$

○, ref. 287; ◐, ref.288; ●, ref.280 +,ref.297 (Theoretical, SSH)

CO_2 (001) + CO ($v=0$) ⟶ CO_2 (000) + CO ($v=1$) + 206 cm^{-1}

□, ref. 289

$N_2$0 (001) + CO ($v=1$) ⟶ $N_2$0 (000) + CO ($v=1$) + 81 cm^{-1}

■, ref. 290

CO ($v=6$) + CO ($v=1$) ⟶ CO ($v=7$) + CO ($v=0$) + 158 cm^{-1}

▼, ref. 291

CO ($v=1$) + CO ($v=1$) ⟶ CO ($v=2$) + CO ($v=0$) + 28 cm^{-1}

△, ref. 291

CO ($v=1$) + COS (000) ⟶ CO ($v=0$) + COS (001) + 81 cm^{-1}

◇, ref. 340

FIG. 4.17. Landau-Teller plots for CO mixtures

The system CO+CH_4 behaves similarly over the short temperature range 303-363 K.[282,294] The other transfers all involve long-range transition dipole interactions, involving a tendency to *negative* temperature-dependence which competes, for non-resonant exothermic processes, with the positive $T^{-\frac{1}{3}}$ dependence expressed by the exponential term in ΔE in eqn (3.3). This is beautifully illustrated by measurements of Powell[291] on the system

$$\text{CO}\,(v{-}1) + \text{CO}\,(v{=}1) \rightarrow \text{CO}\,(v) + \text{CO}\,(v{=}0) + \Delta E.$$

For the case v=2, where ΔE = 28 cm^{-1}, so that the transition is near-resonant, the long-range forces produce a marked negative temperature dependence. In contrast, where v = 7 and ΔE is increased to 158 cm^{-1}, giving a definitely non-resonant transition, the temperature-dependence becomes positive. The system CO_2+CO (ΔE = 206 cm^{-1}) shows a positive dependence for $T > 300$ K, becoming zero between 300 and 200 K, whilst N_2O+CO, with the lower exothermicity, ΔE = 81 cm^{-1}, appears to show a negative temperature dependence at lower temperatures ($T < 200$ K), becoming zero between 200 and 500 K. Transfers between CO (v=1) and the strongly infrared active COS ($00^0 1$), for which ΔH = 81 cm^{-1}, show marked negative dependence between 195 and 370 K.[340] A negative temperature-dependence has also been observed by Richman and Millikan[296] for the system CO + CS_2, and this has been used as an indicator for preferring the multi-quantum near-resonant mechanism

$$\text{CO}\,(v{=}1) + \text{CS}_2\,(000) \rightarrow \text{CO}\,(v{=}0) + \text{CS}_2\,(10^0 1) - 42\ \text{cm}^{-1},$$

to the single-quantum non-resonant mechanism

$$\text{CO}\,(v{=}1) + \text{CS}_2\,(000) \rightarrow \text{CO}\,(v{=}0) + \text{CS}_2\,(00^0 1) \ = \ 608\ \text{cm}^{-1}.$$

CO_2 is a non-polar molecule with two infrared active modes (see p.104). V-V transfers involving the lower frequencies may involve a multiplicity of paths and be difficult to interpret, but reliable data are available for single-quantum transitions involving the ν_3 mode (2349 cm^{-1}). Fig. 4.18 shows a plot of log Z_{AB} against ΔE at 300 K for several mixtures with other infrared active molecules, and with N_2. The infrared-active molecules show the same kind of pattern as the plot for CO mixtures in Fig. 4.16, but with a good deal of scatter. The measurements on CO_2+C_2D_4[316] have been extended over the temperature range 300 to 800 K, and show a consistent *negative* temperature dependence, confirming the importance of interaction between transition dipoles in this almost resonant transition (C_2D_4 (ν_9 = 2345 cm^{-1}) is strongly infrared active). The near-resonant transfer between CO_2 and N_2 (ΔE = 18 cm^{-1}) has been investigated over a wide temperature range. The Landau-Teller plot is given in Fig. 4.19 and shows negative temperature dependence at temperatures below 1000 K, inverting to a marked positive dependence at higher temperatures. The N_2 molecule is neither polar nor infrared active, but Sharma and Brau[314] have explained the inversion as due to a long-range interaction between the transition dipole moment of CO_2 and the *quadrupole moment* of N_2 which is effective for near-resonant transitions only. Quantitative calculations on this basis show fair agreement with the experimental (laser fluorescence) data between 300 and 1000 K. An alternative treatment on similar lines has been given by Andreev.[320] This behaviour is in striking contrast to the system CO+N_2 (ΔE = 188 cm^{-1}), shown in Fig. 4.17, where the smaller transition dipole of CO combined with a larger value of ΔE render the dipole-quadrupole mechanism

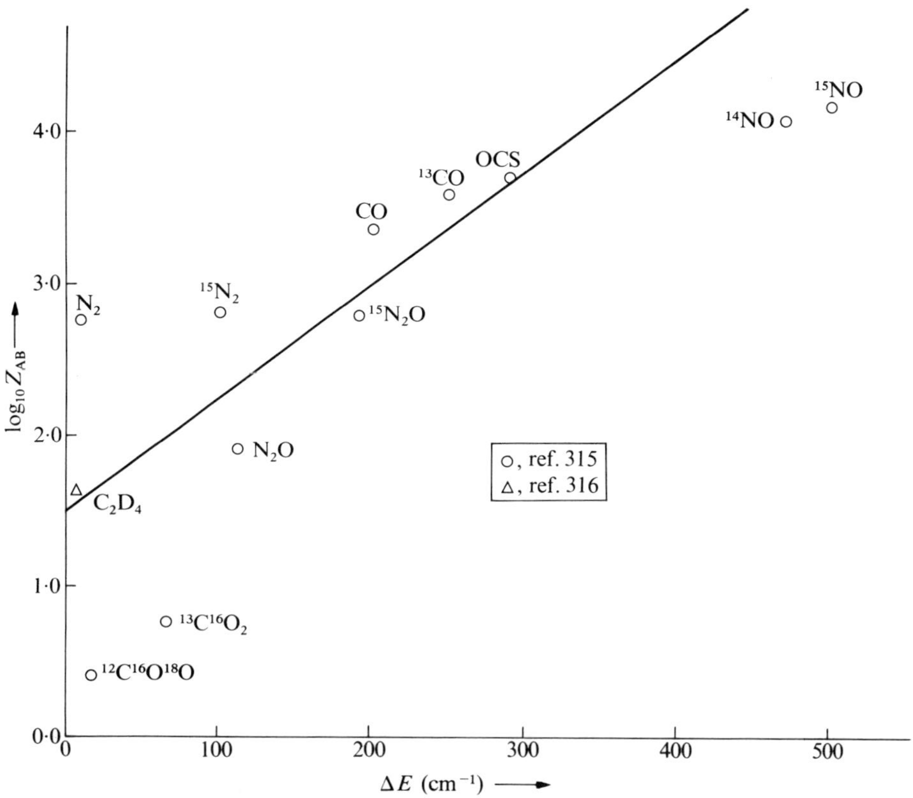

FIG. 4.18. Collision numbers for V-V transfer at 300 K from CO_2 (00^{0}1) in mixtures. Solid line is the slope of Lambert-Salter plot II (Fig. 3.12)

inoperative, so that the normal SSH (positive) temperature-dependence is shown, corresponding to a predominantly repulsive interaction.

A negative temperature-dependence has also been observed for the transition,

$$N_2\,(v{=}1) + N_2O\,(00^0 0) \rightarrow N_2\,(v{=}0) + N_2O\,(00^0 1) + 107\ \text{cm}^{-1},$$

between 300 and 1200 K.[317] A single laser fluorescence experiment gives a value of $Z_{AB} = 1{\cdot}15 \times 10^3$ at 300 K, close to that shown for N_2+CO_2 in Fig. 4.18, and quantitative calculations by the Sharma-Brau theory[314] assuming a dipole-quadrupole interaction again show good agreement with experiment. Theoretical calculations have also been made by Sharma[319] for the transition

$$CO_2\,(00^0 1) + N_2O\,(000) \rightarrow CO_2\,(000) + N_2O\,(00^0 1) + 125{\cdot}8\ \text{cm}^{-1},$$

using a second-order formulation of his dipole-dipole theory, and give close agreement with the experimental value shown in Fig. 4.18. It is suggested that a similar calculation may explain the very high efficiency shown by the transitions involving mixtures of isotopic CO_2 molecules.

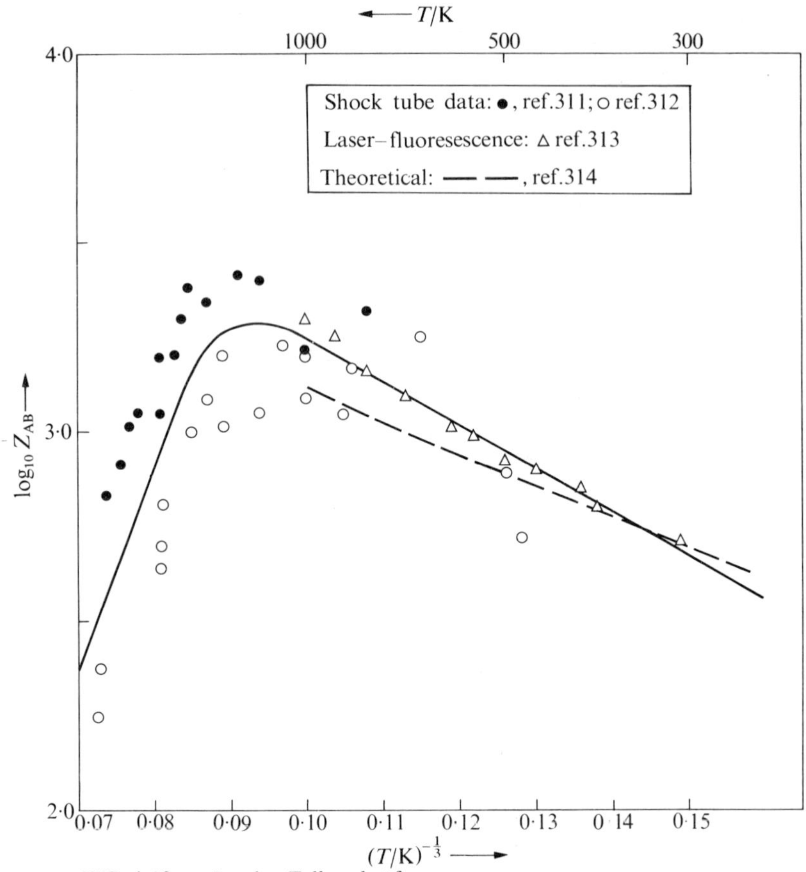

FIG. 4.19. Landau-Teller plot for
$CO_2\,(00^0\,1) + N_2\,(v{=}0) \rightarrow CO_2\,(00^0\,0) + N_2\,(v{=}1) + 19\ \text{cm}^{-1}$

4.4.2. *Polar molecules: H halides*

The hydrogen halides are all strongly polar and infrared active. Their relaxation has been widely investigated by both chemiluminescence and laser fluorescence techniques. The data are summarized on a plot of $\log vZ_{AB}$ against ΔE in Fig. 4.20. It may be noted that this plot extends to much higher values ($\Delta E \leqslant 3000\ \text{cm}^{-1}$) than the corresponding plot for CO mixtures in Fig. 4.16 ($\Delta E \leqslant 800\ \text{cm}^{-1}$), and the relaxation of the halides at corresponding values of ΔE is much faster. This is to be expected, since for most of these mixtures the effect of very strong intermolecular forces, due to the static dipoles, will be added to the effect due to the interaction of the transition dipoles. The halides also possess low moments of inertia (see Table 3.3), so that V-R transfer, simultaneous with V-V transfer, is likely to play a significant role. The data in Fig. 4.20 show systematic regularities:

(a) There is little distinction between the relaxation efficiencies of systems involving HCl and HBr. In contrast, relaxation efficiencies of systems involving

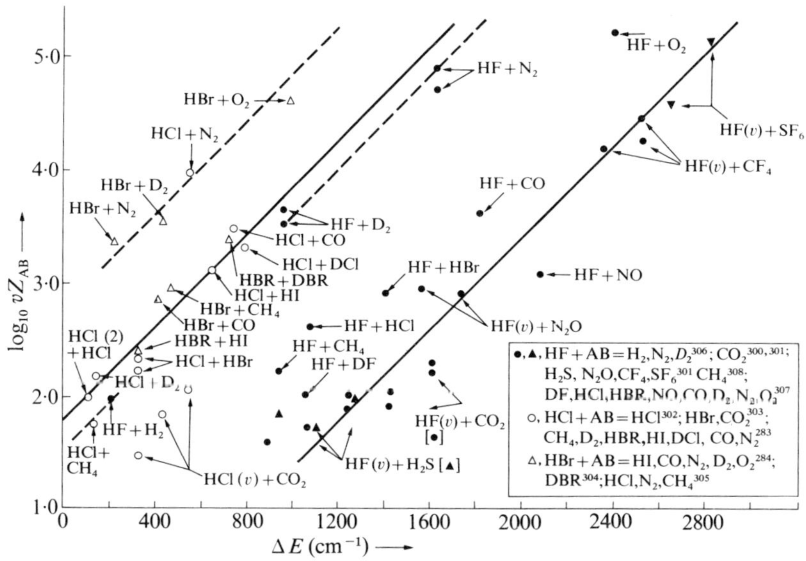

FIG. 4.20. Reduced collision numbers for V-V transfer at 300 K
$HX(v) + AB(0) \leftrightarrow HX(v-1) + AB(v=1) \pm \Delta E$.
All values calculated for transition in the exothermic direction. Transitions are for $HX(v=1)$ unless labelled $HX(v)$. Solid lines, transitions to infrared-active modes. Dashed lines, transitions to infrared-inactive modes

HF are roughly two orders of magnitude higher. It is interesting to compare Fig. 4.20 with Fig. 3.7, which shows the same pattern of behaviour for V-T transfer. Table 3.3 shows that the dipole moment of HF is approximately twice that of HCl and the moment of inertia approximately half that of HCl, while the differences between HCl and HBr are much less pronounced. These differences are clearly reflected in the data for both V-T and V-V relaxation.

(b) For each group, relaxation in mixtures with infrared-inactive collision partners, H_2, D_2, N_2, O_2 is one or two orders of magnitude slower than relaxation involving a second infrared-active mode (CO, NO, CH_4 (ν_3), H_2S (ν_3), CF_4 (ν_3), SF_6 (ν_3), N_2O (ν_3), and the other halides). Relaxation to the strongly infrared-active CO_2 (ν_3) appears to occur with an anomalously high efficiency for both groups. An anomalously high efficiency is also shown by the mixture HCl ($v=2$) + D_2, which relaxes at almost the same rate as HCl ($v=2$) + HCl. This transition is near-resonant (ΔE = 108 cm^{-1}), and it is possible that interaction of the transition dipole of HCl with the quadrupole moment of D_2 plays a role similar to that discussed in § 4.4.1 for mixtures of N_2 with CO_2 and N_2O.

(c) Each group shows roughly the same linear dependence of log Z_{AB} on ΔE. In terms of eqn (3.12) this corresponds to a value of the slope $C = 5{\cdot}3$ cm, which is significantly smaller than that of the Lambert-Salter plot I for hydrogen-containing molecules, $C = 8{\cdot}3$ cm.

(d) The 'horizontal scatter' (cf. Fig. 3.14) of the HF data amounts to roughly ±200-±300 cm^{-1}, which can be attributed to simultaneous V-R transfer (cf. § 3.4.3). The majority of HF molecules at 300 K will be in the J=1 or J=2 levels, and a transition $\Delta J = \pm 2$ would involve a contribution $\Delta E \sim 200$ or ~ 280 cm^{-1} to the overall transition. The scatter of points for HCl and HBr is less than would be expected on similar grounds.

The general behaviour of the hydrogen halides is in qualitative accord with the theoretical trends discussed in § 4.4.1 for CO mixtures, and this extends to temperature dependence. HF mixtures have been investigated over the temperature range 200 to 1000 K by Bott and Cohen.[307] They find the expected positive Landau-Teller dependence for mixtures with D_2 and H_2. In contrast, mixtures with HCl, DF, and CO_2 show a negative dependence at lower temperatures, which inverts to a positive dependence between 700 and 800 K. For HCl and HBr the process

$$\text{HX}\ (v{=}1) + \text{HX}\ (v{=}1) \rightarrow \text{HX}\ (v{=}2) + \text{HX}\ (v{=}0) + \Delta E.$$

has been investigated by Noter *et al.*[321] and both show a consistently negative temperature dependence between 300 and 700 K. Quantum-mechanical calculations, involving multi-quantum rotational transitions simultaneous with the V-V transition, have been made by Dillon and Stephenson[309] for the highly efficient transfers between HF or HCl and CO_2. These show better agreement with experiment than the simple dipole-dipole theory of Sharma,[293] which underestimates the efficiency.

4.5 Mechanisms of vibrational excitation

The vibrational excitation of a gas may involve both V-V and V-T transfer. The general principles are well illustrated by the case of CO, for which detailed information about both kinds of process is available. Comparison of Fig. 3.3 (V-T transfer) with Fig. 4.16 (V-V transfer) shows that at 300 K near-resonant V-V transfer ($Z \sim 3 \times 10^2$) is several orders of magnitude faster than V-T transfer ($Z \sim 10^{10}$). V-T transfer shows a strong positive temperature dependence, and V-V transfer a weak negative temperature dependence (Fig. 4.17), but even at 1000 K V-T transfer ($Z \sim 10^{6{\cdot}6}$) is still at least three orders of magnitude slower. The mechanism for vibrational excitation of cold CO (v=0) to upper vibrational levels will therefore be an initial V-T excitation from (v=0) to (v=1), followed by successive V-V transfers to higher levels. Since transitions involving $\Delta v = \pm 1$ are the most efficient, this will be a stepwise process:

$$2\text{CO}\ (v{=}0) + 2143\ \text{cm}^{-1} \rightleftharpoons \text{CO}\ (v{=}1) + \text{CO}\ (v{=}0) \qquad \text{V-T} \qquad (4.3)$$

$$2\text{CO}\ (v{=}1) \underset{k_2}{\overset{k_1}{\rightleftharpoons}} \text{CO}\ (v{=}2) + \text{CO}\ (v{=}0) + 26{\cdot}3\ \text{cm}^{-1} \qquad \text{V-V} \qquad (4.4)$$

$$2CO\ (v{=}2) \underset{k_2}{\overset{k_1}{\rightleftharpoons}} CO\ (v{=}3) + CO\ (v{=}1) + 26{\cdot}3\ \text{cm}^{-1} \qquad \text{V-V} \qquad (4.5)$$

It will be noted that, owing to anharmonicity, processes (4.4) and (4.5) are not exactly resonant (as they would be for a harmonic oscillator). Steps involving more widely separated levels, e.g.

$$CO\ (v{=}3) + CO\ (v{=}1) \rightleftharpoons CO\ (v{=}4) + CO\ (v{=}0) + 79\ \text{cm}^{-1},$$

become less closely resonant, and will be slower. The overall vibrational excitation process will thus be that represented diagrammatically in Fig. 4.21.

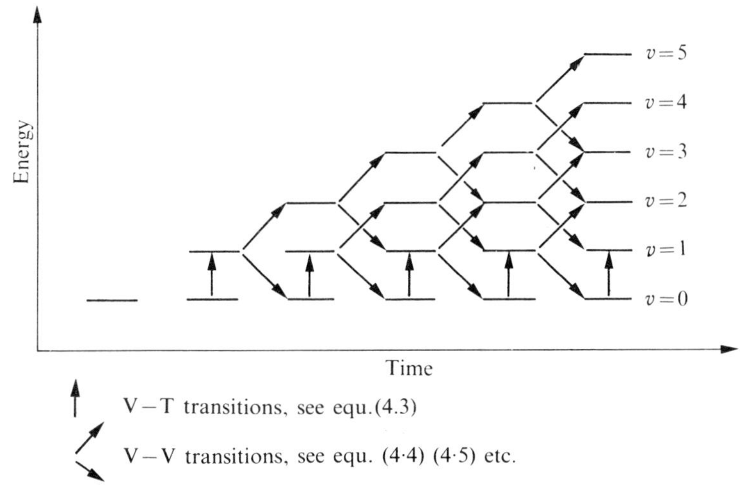

FIG. 4.21. Mechanism for vibrational excitation

The slowest step will be V-T activation to the (v=1) level, which will be rate-determining, and the subsequent V-V transfers will occur at increasingly fast rates with increasing values of v ($P_{v,v-1} - vP_{10}$). A mechanism of this kind has been confirmed by shock-tube experiments on mixtures of 10 per cent HI with 90 per cent Ar, in which the populations of HI molecules in levels v = 1, 2, and 3 were followed spectroscopically.[328] The conditions of the experiment were equivalent to the HI, initially at 300 K, being suddenly plunged into a heat bath of inert gas at 2000 K, and it was found that all three vibrational levels were produced at about the same rate, i.e. the rate-determining V-T transfer to v=1. In deactivation from higher vibrational temperatures the reverse process will occur. There will be rapid stepwise V-V transfer until the v=1 level is reached. The molecules will accumulate in this level, producing an overpopulation which may amount to a population inversion, and will relax slowly to v=0 by V-T transfer. This mechanism was originally proposed by Norrish[329] to explain the observation that in flash photolysis experiments producing NO ($v > 1$) the relaxation of the upper levels is too fast to be observed by kinetic spectroscopy, but the slower relaxation of NO (v=1) can be easily followed.

Similar conditions will apply for thermally activated chemical reactions, where a large proportion of molecules will be in higher vibrational levels, so that critical activation of deactivation will involve mainly V-V transfer. For simple unimolecular isomerization or decomposition reactions, e.g.

$$c\text{-}C_3H_6 \rightarrow CH_3CH{:}CH_2$$

and

$$c\text{-}C_4H_8 \rightarrow 2C_2H_4,$$

the kinetics are first order only at pressures where a Boltzmann distribution of chemically activated molecules is maintained by collisional activation: at lower pressures the reaction becomes second order. The relative efficiencies of collisions with different chemically inert additive molecules may be investigated by following their effect on the pressure at which the change from first- to second-order kinetics occurs. The two reactions listed above have been investigated in this way by Pritchard *et al.*[338,339] and their results are summarized in Table 4.4.

TABLE 4.4.

Relative collision efficiencies of gases in energy transfers with reacting molecules in unimolecular reactions (data from refs. 338 *and* 339)

Additive	Reactant Temperature	$c\text{-}C_3H_6$ 492 K	$c\text{-}C_4H_8$ 448 K
Self		1	1
He		0·05	0·07
H_2		0·12	0·10
Ne		–	0·12
Ar		0·07	0·21
N_2		0·07	0·21
CH_4		0·24	0·38
H_2O		0·74	0·44
$C_6H_5CH_3$		1·10	1·12
$C_6H_3(CH_3)_3$		0·89	1·23

The most efficient collision partners are complex organic molecules, which have a wide spectrum of vibrational frequencies so that rapid near-resonant V-V transfer can occur with the reactant. These show collision efficiencies as great as, or slightly larger than, self-collisions of the reactant. In contrast, the smaller and lighter molecules, He, H_2 etc., are very inefficient, in spite of their known efficiency as collision partners for V-T transfer (see Table 3.10). The inference follows that V-T transfer does not play an important role.

The mechanism discussed above leads to interesting conclusions about the evolution of the distribution of molecules among the vibrational levels during a relaxation. For harmonic oscillators processes (4.4) and (4.5) would be exactly resonant, so that the rates in both directions, k_1 and k_2, will be equal. Under these conditions Montroll and Shuler have shown that an initial Boltzmann distribution will relax to a final equilibrium by a continuous series of Boltzmann distributions.[330] This applies both to relaxation in a heat-bath at constant translational temperature, and in an isolated ensemble of oscillators.[331] For anharmonic oscillators, in contrast, processes (4.4) and (4.5) are non-resonant (as for CO) and the rates in the exothermic and endothermic directions will differ:

$$k_1(\text{exo}) = k_2(\text{endo}) \exp(\Delta E/RT).$$

The equilibrium constant for the reversible process will no longer be unity, but will favour the less symmetrical distribution, corresponding to a higher and a lower value of v. This will give rise to a marked departure from Boltzmann vibrational distribution during relaxation.

A detailed study by Treanor[332] has shown that the distortion from Boltzmann distribution becomes most marked under conditions of low translational temperature and high vibrational energy. Under these conditions there will be a progressive build-up of population in the higher levels, which can amount to a population inversion. This build-up is limited at *very* high levels because V-T transfer, which becomes much faster at these levels due to anharmonicity, begins to play a competing role. The efficiency of V-T transfer varies exponentially with ΔE (eqn (3.3)), so that the V-T transfer,

$$\text{CO}\,(v{=}41) + \text{CO}\,(v{=}0) \leftrightarrow \text{CO}\,(v{=}40) + \text{CO}\,(v{=}0) + 1050\,\text{cm}^{-1},$$

involves an exothermicity approximately *half* that for the $0 \rightarrow 1$ excitation (eqn (4.3)). This, coupled with the linear variation $P_{v,v-1} = vP_{1,0}$ corresponds roughly to a value $Z \sim 10^2$ at 1000 K, which is comparable with the near-resonant V-V efficiency. V-T relaxation, in contrast to anharmonic V-V relaxation, occurs *via* a series of Boltzmann distributions.

The overall picture is shown in Fig. 4.22, which is based on calculations by Bray[333] for CO. This is a logarithmic plot of X_v, the population of the level v relative to the level (v=0), against the energy of the vibrational level, expressed as E_v/D where D is the dissociation energy of the molecule, (the corresponding values of v are shown at the top of the diagram). There are four plots on the diagram:

(a) The dotted straight line represents a Boltzmann distribution corresponding to vibrational and translational temperatures, $T_{TR} = T_{VIB} = 3000$ K.

(b) The upper (roughly parabolic) chain curve represents the distorted equilibrium predicted by Treanor for exclusively V-V transfer, where $T_{VIB} = 3000$ K and $T_{TR} = 800$ K: it rises to $X_v \sim 1$ at v=1 and at $v > 40$, and shows a pronounced minimum at $v \sim 20$.

(c) The dashed straight line represents the Boltzmann distribution produced by V-T relaxation from levels where $v > 40$, under the same temperature conditions as (b).

(d) The actual distribution, represented by the full curve, is made up by

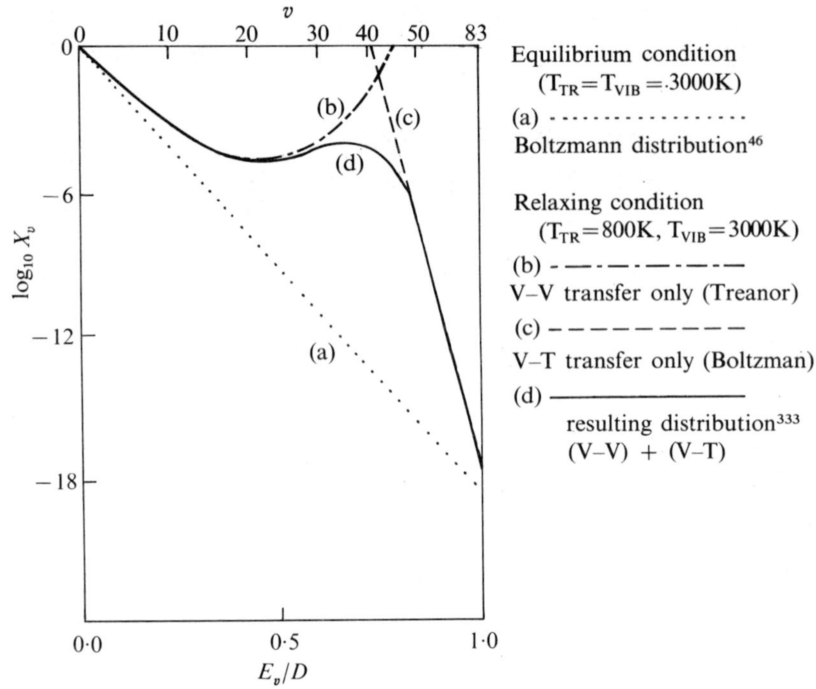

FIG. 4.22. Vibrational population distributions for CO

a combination of plots (b) and (c). This deviates strongly from the Boltzmann distribution (a), and shows a distinct population inversion in the region $v = 20$-30. This is of importance in the operation of CO lasers; the 'kinetic modelling of high-power CO lasers' has been investigated by Rich.[334]

Disturbed equilibrium of vibrational populations is also shown in binary mixtures of diatomic gases where non-resonant V-V transfer occurs between the two components, for example, the reversible process

$$N_2\,(v{=}1) + CO\,(v{=}0) \underset{k_2}{\overset{k_1}{\rightleftharpoons}} N_2\,(v{=}0) + CO\,(v{=}1) + 187\ \text{cm}^{-1}.$$

Here k_1 (exothermic) will be faster than k_2 (endothermic), and this will lead to a pumping of the vibrational energy of the system into CO, which will show an apparently higher vibrational temperature. This is important in the operation of the CO-N_2 laser, and has been investigated experimentally by Legay *et al.*[335] It also plays a role in the nozzle expansion of compressed N_2+CO mixtures, which has been investigated by Teare, Taylor, and von Rosenberg.[336]

5

ROTATION–TRANSLATION TRANSFER

5.1. Theoretical treatments

5.1.1. General survey

The general characteristics of R-T transfer, and how these differ in important respects from those of V-T transfer, were discussed in §§ 1.3.2 and 1.5. Detailed theoretical treatments of rotationally inelastic collisions fall into three classes—quantum mechanical, classical, and semiclassical. The area of application of each method depends on the magnitudes of the rotational and translational quantum numbers involved. When both rotational and translational quantum numbers are small, as in collisions involving H_2 and He, the quantum method is applicable. For systems specified by low rotational and high translational quantum numbers, e.g. HCl + Ar, the semiclassical method may be applied. For the very large number of systems involving heavy molecules with large moments of inertia, where both quantum numbers are large, the classical method is the most suitable. Calculations based on all three methods require an accurate intermolecular potential, $V = V(r, \theta)$, particularly its angular dependence, which is only available for a few simple systems. Reviews of the theoretical treatment of R-T transfer have been published by Takayanagi in 1965[84] and in 1973,[370] and by Secrest in 1973.[86]

5.1.2. Classical theories

The first classical treatment of rotational transfer was due to Parker,[21] who investigated collisions between two homonuclear diatomic molecules, AB and CD. He adopted a potential function

$$V = a\,[(\exp(-\alpha r_{AC}) + \exp(-\alpha r_{AD}) + \exp(-\alpha r_{BC}) + \exp(-\alpha r_{BD})] \\ - b\,[\exp(-\tfrac{1}{2}\alpha R)],$$

containing both repulsive and attractive terms, and confined his attention to coplanar encounters with the initial rotational energy zero. Solution of the classical equations of motion showed that rotational relaxation could be characterized by a single relaxation time, β_R, involving the whole of the rotational energy of the molecule. This leads to a rotational collision number, Z_R, given by

$$Z_R = \frac{Z_R^{\infty}}{\left[1 + \frac{\pi^{\frac{3}{2}}}{2}\left(\frac{T^*}{T}\right)^{\frac{1}{2}} + \left(\frac{\pi^2}{4} + \pi\right)\left(\frac{T^*}{T}\right)\right]}, \qquad (5.1)$$

where $Z_R = \frac{1}{16}(\alpha d/\epsilon)^2$ and $T^* = \epsilon/k$. (α, d, and ϵ are the intermolecular potential parameters). Eqn (5.1) predicts a slight positive temperature-dependence for Z_R (transfer becomes less efficient at higher temperatures and speeds of rotation), and gives fair predictions for O_2 and N_2 at 300 K (see Table 5.2). The Parker theory has been elaborated by Brau and Jonkman.[388]

Widom[365] used a classical 'rough sphere' model for the rotational relaxation of a spherical-top molecule in collision with an inert gas molecule, and found the simple expression

$$Z_R = \tfrac{3}{8}(1+b)^2/b, \tag{5.2}$$

where $b = I/\mu a^2$, I being the moment of inertia, μ the reduced mass of the collision, and a the molecular diameter. Eqn (5.2) predicts a *minimum* value $Z_R = 1{\cdot}5$ when $I = \mu a^2$ and $b = 1$. It will be seen from Table 5.4 that such a minimum does occur for N_2, where collisions with Ne (M=20) are more efficient than those with either He (M=4) or Ar (M=40), also for p-H_2 where classical theory does not apply. The Widom theory gives fair quantitative predictions for mixtures of CH_4 and CD_4 with inert gases, which do not show the minimum condition. Eqn (5.2) is analogous to an expression obtained earlier by Wang Chang and Uhlenbeck,[366] who applied the same model to rotational relaxation of spherical-top molecules in self-collisions, and obtained a similar expression:

$$Z_R = \tfrac{3}{8}(1+2b)^2/b. \tag{5.3}$$

Predictions for pure CH_4 and CD_4 are compared with experiment in Table 5.2. Both of these models neglect attractive forces entirely, and predict values of Z_R which are independent of temperature.

Sather and Dahler[367] also made calculations for rotational relaxation of perfectly rough spheres in collision with atoms, and obtained an expression for Z_R identical with (5.2), but multiplied by an additional term $(m_A + m_B)/m_A$, where m_A is the mass of the spherical molecule and m_B that of the atom. This gives an increased efficiency for collision with lighter atoms. The difference is due to Widom having assumed the rough sphere to be initially at rest in deriving (5.2). Sather and Dahler also derive an expression for relaxation of smooth, solid, prolate sphero-cylinders in self-collision, and find that the efficiency of energy transfer depends mainly on a parameter

$$\alpha = mL^2/8(I_A I_B I_C)^{\frac{1}{3}},$$

where m is the mass of the molecule, L the length of the cylinder and I_A, I_B, I_C the three moments of inertia. In a later paper[368] the treatment is extended to include attractive forces by using a square-well potential. This gives rise to a positive temperature-dependence for Z_R which is stronger than that predicted by Parker[21] for molecules such as N_2.

More recently it has become possible to make detailed classical trajectory calculations for suitable systems, followed by a Monte Carlo averaging procedure to evaluate a rotational relaxation time and collision number.

Kistemaker and de Vries[361] have made calculations for mixtures of N_2 with inert gases, assuming an exponential Morse potential between the individual atoms, with some adjustments for anisotropy. Their results show fairly good agreement with experiment (Table 5.4), in spite of the approximations involved in their intermolecular potential, and it is interesting that they predict the *minimum* value of Z_R for the mixture with Ne which is observed experimentally and was predicted by Widom. Their paper reviews a number of earlier trajectory calculations.

5.1.3. *Quantum-mechanical theory*

Quantum theories, in contrast to classical, evaluate the probability of transition between *individual* rotational levels J and J'. The treatment is in essentials the same as that outlined for V-T transitions in § 3.1.1. Early papers by Beckerle[371] and Brout[372] used the distorted wave approximation, while Takayanagi[373,374] introduced the modified wave number approximation. A further analysis was made by Davison.[375,376] Recently numerical calculations based on these methods have been made by Agrawal and Saksena.[377,378,380] The treatment of rotational transitions requires an angle-dependent exponential molecular potential of the form

$$V(r, \theta) = V_1(r) + \beta \exp(-\alpha r) \cos^2 \theta.$$

This is more complicated than the potential (3.1) used in the SSH treatment of V-T transfer, and the asymmetry factor β must usually be treated as an adjustable parameter in default of more accurate information. The theory predicts that the value of Z_R for an individual rotational transition $J \rightarrow J'$ will increase rapidly with increasing ΔE and will decrease with rising temperature, in analogy with V-T and V-V transfer.

An approximate treatment for heavier homonuclear diatomic molecules was developed by Brout,[379] who found a mean R-T transfer probability given by the simple, temperature-independent relation $P = \frac{1}{2}(d_0/r_0)^2$, where d_0 is the internuclear distance in the molecule, and r_0 the kinetic collision diameter. The reasons behind the simplicity of this result are interesting. Temperature disappears from the relation because the increased spread of occupied rotational levels at higher temperature causes a lowering of the probability, which cancels out the favourable effect of increased velocity of approach. Mass disappears because a larger moment of inertia brings the rotational levels closer together and increases the probability, cancelling out the adverse effect of the decreased velocity of approach of the heavier molecules. The intermolecular potential is unimportant, as the collision time is always short with respect to the period of the rotation. The theory predicts values of $Z_R = 17$ for O_2 and 20 for N_2, which are three to four times the experimental values at 300 K (Table 5.2), and are clearly not quantitatively reliable.

5.1.4. Semiclassical theory

These theories use a classical approximation to the translational motion in a collision, combined with a quantum treatment of the rotational motion. A detailed treatment of the HCl+Ar system has recently been made by Neilsen and Gordon,[189] whose procedure is best described by quoting from their Introduction: 'The present paper describes a method of calculating the relation between potential and sigma matrix for an atom-diatom collision, using a semiclassical approximation to the collision dynamics. The translational motion of the atom-diatom system is specified classically, in terms of the intermolecular distance and the angle between the original and current direction of the intermolecular axis. The classical equations of motion are used to obtain these quantities as a function of time. The rotational motion of the diatom is specified by the quantum numbers J, the rotational angular momentum, and m, its projection along the intermolecular axis. The time-dependent Schrödinger equation is used to relate the rotational state of the system before a collision to the rotational state afterwards. The methods of averaging over initial conditions and calculating experimental quantities using the impact approximation are also described.' There are no experimental measurements yet available for comparison, but it is interesting that these accurate computations confirm the positive temperature-dependence of Z_R predicted by the simple classical treatments.

5.1.5. Comparison with experiment

Rotational relaxation is less easy to investigate by acoustic measurements than vibrational relaxation, as collision numbers and relaxation times are much smaller and measurements must be made at high f/p values where large corrections for classical absorption and dispersion are needed (see § 2.2.1). A further difficulty is that, while heavy molecules possessing large moments of inertia always show a single relaxation time in accord with theory, lighter molecules will show multiple relaxation processes: e.g. p-H_2 which shows, in collision with He at 300 K, two relaxation times corresponding to the $J = 0 \rightarrow 2$ and $J = 2 \rightarrow 4$ transitions. Such relaxation times lie close together and the situation is not always easy to analyse satisfactorily. Nevertheless the acoustic method provides the only reliable technique for measuring rotational relaxation times, except in the rare cases where spectroscopic techniques may be used, and all the experimental results given in Tables 5.1, 5.2, 5.3, and 5.4 have been taken from acoustic sources. Rotational collision numbers obtained from measurements of the transport properties are highly suspect (see § 2.9), and none are quoted.

5.2. Relaxation in pure gases

5.2.1. Light diatomic molecules: H_2, D_2, HD

n- n-H_2 is a mixture of three parts o-H_2 to one part p-H_2. At 300 K nearly 90 per cent of the o-H_2 will be in the J=1 level, while approximately 50 per cent of the p-H_2 will be in the J=0 level and 50 per cent in the J=2 level. Only transitions involving $\Delta J = \pm 2$ are allowed for either modification. The processes involved in rotational relaxation at 300 K may be summarized as follows:[381]

	$J-J'$	population (per cent)	ΔE (cm^{-1})
p-H_2	0 – 2	18·7	353
	2 – 4	9·4	826
	4 – 6	0·2	1294
o-H_2	1 – 3	66·4	589
	3 – 5	5·2	1062
	5 – 7	—	1530

If the temperature is lowered to 100 K, virtually the whole of the p-H_2 will be in the J=0 level, and the whole of the o-H_2 in the J=1 level. For D_2 (which has a nucleus containing an even number of particles) the roles of the ortho and para forms are interchanged and the rotational quantum differences are smaller due to the larger moment of inertia, but at temperatures below 50 K o-D_2 will be exclusively in the J=0 level, and p-D_2 in the J=1 level. (n-D_2 consists of two parts o-D_2 to one part p-D_2). Relaxation measurements made on the pure ortho and para modifications at lower temperatures will thus be concerned entirely with J=1→3 or J=0→2 transitions, and provide an excellent test of the quantum-mechanical theories.

The results of acoustic experiments are given in Table 5.1, and show fair agreement with theoretical prediction. The Z_R values for p-H_2, where the J=0→2 transition (ΔE = 353 cm^{-1}) is mainly involved, decrease consistently with rising temperature from Z_R = 717 at 77 K to Z_R = 411 at 170 K. The value for o-H_2 at 170 K, corresponding to the J=1→3 transfer with the higher value of ΔE = 589 cm^{-1}, is Z_R = 790, nearly double the value for p-H_2 at the same temperature. This falls further to Z_R = 510 at 233 K, but remains constant between 233 and 269 K, where the more energetic J=3→5 transition (ΔE = 1062 cm^{-1}) will begin to play an appreciable role, and then rises to Z_R = 530 at 293 K.

The relaxation of n-H_2 at 300 K will involve two main transitions, J=1→3 for ortho and J=0→2 for para-H_2. The published reports of acoustic measurements all show absorption and dispersion regions extending over larger ranges of f/p than predicted for a single relaxation time, but satisfactory resolution of the individual times has not been possible.[341,342,387] The values of Z_R given in Table 5.1 are the 'best fit' to a single relaxation time. Winter and Hill[342] made a rough estimate for the two separate processes in n-H_2 at 295 K, leading to the values Z_R (J=0→2) = 191 and Z_R (J=1→3) = 278. These values are much lower than would be anticipated from the results for pure o-H_2 and p-H_2, which suggests that collisions between p-H_2 and o-H_2 are more efficient than self-collisions of either modification. This is confirmed by the measurements of Jonkman *et al.*[344] on an artificial mixture of p-H_2 with o-H_2 at 77 K. At this

TABLE 5.1.

Rotational relaxation in H_2, D_2, HD

Molecule	Temp. (K)	Z_R (exp.)	Ref.	Z_R (calc.)	Method*	Ref.
p-H_2	77	720	389	–	–	–
	77	717	343	–		
	77·3	715	344	–		
	90	628	343	–		
	90·5	635	344	–		
	115	604	344	–		
	170	411	344	–		
	300	–	–	$Z_{20} = 329$	DW	372
o-H_2	170	790	345			
	233	510	345			
	269	410	345			
	293	530	345			
	300	–	–	$Z_{31} = 338$	DW	372
n-H_2	77	535	389	–		
	273	350	341	–		
	295	220	342	–		
	873	303	342	–		
	1073	307	342	–		
p-H_2* + *o*-H_2	77	394	344	–		
				–		
o-D_2	31·2	370	346	–		
	40·6	381	346	–		
	44·5	413	346	–		
	52·5	399	346	–		
	77	296	343	–		
	77·3	308	346	–		
	90	232	343	–		
	90·5	268	346	–		
n-D_2	295	188	342	–		
	773	225	342	–		
	1073	328	342	–		
HD	20·0	–	–	25	DW	382
	20·4	11·1	347	–		
	27·0	10·6	347	–		
	32·1	10·4	347	–		
	39·6	11·2	347	–		
	39·9	10·1	347	–		
	40·0	–	–	19	DW	382
	42·6	10·6	347	–		
	293	<30	343	–		

*DW = quantum-mechanical distorted wave calculations

temperature the *o*-H_2 will be exclusively in the $J=1$ level and there is insufficient energy to promote a $J=1\rightarrow3$ transfer. The observed collision number, $Z_R = 394$, will thus correspond to the process

$$p\text{-H}_2(J{=}0) + o\text{-H}_2(J{=}1) \rightarrow p\text{-H}_2(J{=}2) + o\text{-H}_2(J{=}1) - 353 \text{ cm}^{-1},$$

which involves a *rotating* collision partner o-H_2 (J=1), as contrasted with the process

$$p\text{-H}_2(J{=}0) + p\text{-H}_2(J{=}0) \rightarrow p\text{-H}_2(J{=}2) + p\text{-H}_2(J{=}0) - 353 \text{ cm}^{-1},$$

which involves a *non-rotating* collision partner, p-H_2 (J=0), and is much less efficient, with $Z_R = 717$. Crawford[390] has made a theoretical investigation using the quantum-mechanical distorted wave approximation, and has shown that the inelastic collision cross-section for rotational transitions depends on the rotational states of *both* colliding molecules. He also predicts that simultaneous R-R transitions, such as

$$p\text{-H}_2(J{=}2) + p\text{-H}_2(J{=}2) \rightarrow p\text{-H}_2(J{=}0) + p\text{-H}_2(J{=}4) - 473 \text{ cm}^{-1}, \quad (5.4)$$

will play an appreciable role at higher temperatures, when sufficient energy is available. But it must be noted that such transitions, in contrast to similar V-V transitions (eqns (4.4) and (4.5)), are both strongly non-resonant and *endo-thermic*, and will be much less efficient.

The accurate measurements reported for HD are all at temperatures below 40 K, where only the J=0→1 transition can occur. This corresponds to an energy $\Delta E = 91$ cm^{-1}, which is about $\frac{1}{4}$ the value for J=0→2 transitions in p-H_2, and leads to the much higher transition probability. The effective constancy of the low value, $Z_R \sim 10$, between 20 and 40 K is attributed by Takayanagi[370,382] to the fact that, at very low molecular velocities, the acceleration produced by the attractive part of the intermolecular potential is the main factor contributing to the transition probability. A similar explanation was put forward in § 3.3.1 for the constancy of Z_{10} for V-T transfer in H_2 at temperatures below 50 K (cf. Fig. 3.6). The approximate estimate of $Z_R \sim 30$ at the higher temperature 293 K is in accord with the expectation that transitions between higher rotational levels will be involved.

5.2.2. *Heavy diatomic molecules:* N_2, O_2, CO, NO, HCl

The four molecules, N_2, O_2, CO, and NO all possess large moments of inertia giving spectroscopic constants $B < 2$ cm^{-1}. R-T transfer is accordingly much more efficient, and the majority of measurements give values of $Z_R \leqslant 5$. All will occupy a wide spectrum of higher J levels, and may be treated by classical theory, which predicts a single relaxation time for the whole rotational energy. Experimental data obtained from acoustic measurements are listed in Table 5.2. The room temperature values for N_2 ($Z_R \sim 5$) and O_2 ($Z_R \sim 4{\cdot}5$) lie close together, and are in fair accord with the calculations of Parker. Both show a steady increase in Z_R with rising temperature as higher rotational levels become active. It is interesting that there is no significant difference between $^{14}N_2$, for which a $\Delta J = \pm 2$ selection rule will apply, and $^{14}N^{15}N$ for which $\Delta J = \pm 1$ transitions are allowed, and it may be inferred that the size of invididual quantum jumps is unimportant. CO and NO both show significantly lower

TABLE 5.2.

Rotational relaxation in heavy diatomic molecules

Molecules	B (cm^{-1})	Temp. (K)	Z_R (exp.)	Ref.	Z_R (calc.)	Method*	Ref.
HCl	10·6	300	7	350	10·5	DW	378
		300	–	–	19	Cl	383
N_2	2·01	77·1	1·8	359	–		
		180	3·9	359	–		
		260	5·1	359	–		
		293	5·6	359	–		
		300	4·7	354	4·01	Cl	21
		300	5·3	355	–		
		773	11·3	342	–		
		1073	15·4	342	–		
$^{14}N^{14}N$		300	3·9	358	–		
$^{14}N^{15}N$		300	4·1	358	–		
CO	1·93	77·1	1·1	359	–		
		180	2·6	359	–		
		260	3·2	359	–		
		293	4·0	359	–		
		300	2·1	357	–		
		300	2·8	358	–		
NO	1·7	300	~1·0	144	–		
O_2	1·45	300	4·1	354	3·45	Cl	21
		300	4·1	355	–		
		300	4·5	356	–		
		773	12·9	342	–		
		1073	12·9	342	–		
		1773	11·6	342	–		

*DW = quantum-mechanical distorted wave calculation
Cl = classical calculation (Parker)

values of Z_R. CO is non-polar and NO only very weakly polar, so that dipole-dipole interaction seems unlikely to be important, and the possible influence of a $\Delta J = \pm 1$ selection rule has been discounted by the $^{14}N^{15}N$ result. In heteronuclear diatomic molecules the centre of mass does not coincide either with the geometrical midpoint or with the centre of charge-distribution. This will have some effect on the anisotropy of the potential, and may contribute to the low values of Z_R found for CO and NO (also perhaps for HD).

HCl, which shows a large spectroscopic constant, $B = 10{\cdot}6$ cm^{-1}, combined with a very low value, $Z_R = 7$, at 300 K, is strongly polar and is a clear case of dipole-dipole interaction increasing the transfer probability. Quantum-mechanical calculation by Agrawal and Saksena[378] for the $J = 0 \rightarrow 1$ transition predict a higher value, $Z_R = 10{\cdot}9$, while classical calculations on molecular dynamics by Zeleznik[383] give $Z_R = 19{\cdot}0$ and are clearly unsuitable for this case.

5.2.3. *Polyatomic molecules*

Experimental data from acoustic experiments are listed, in order of decreasing value of the rotational constant, B, in Table 5.3. The four polar hydrides H_2O, HCl, NH_3, and H_2S show clear evidence of the effect of dipole-dipole interaction, giving low values of Z_R in spite of the large rotational constants.

TABLE 5.3.

Rotational relaxation in polyatomic molecules

Molecule	B (cm^{-1})	Temp. (K)	Z_R (exp.)	Ref.	Z_R (calc.)	Method*	Ref.
H_2O	15·5	323	4·0	161	3·0	DW	378
NH_3	8·5	300	4·9	177	14·4	Cl (WCU)	177
		300	–	–	1·4	DW	378
H_2S	7·2	300	31	351	14·1	DW	378
				342			
CH_4	5·25	77·1	3·0	342	–		
		180	5·2	342	–		
		260	7·4	342	–		
		293	8·6	342	–		
		300	12	352	–		
		300	14 to 17	353	18	Cl (WCU)	365
CD_4	2·64	300	7	352	12	Cl (WCU)	365
C_2H_4	3·98	300	1·5	352	3·5	Cl (WCU)	352
c-C_3H_6	0·67	300	5	352	10·3	Cl (WCU)	352
$C(CH_3)_4$		300	3·7	178	7·9	Cl (SD)	178
c-C_5H_{10}		300	2·4	178	8·7	Cl (SD)	178
n-C_3H_8		300	1·3	174	3·0	Cl (SD)	174
CO_2	0·39	300	2·5	360	–		
N_2O	0·42	300	2·5	360	–		

*DW = quantum-mechanical distorted wave calculation
Cl = classical calculation: (WCU) – Wang Chang and Uhlenbeck, eqn. (5.3), (SD) – Sather and Dahler

H_2O shows fair agreement with distorted wave calculations by Agrawal and Saksena:[378] the value $Z_R = 4·9$ for NH_3, lies between their prediction, 1.4, and a classical estimate, 14, calculated from eqn (5.3).[177] Estimate of the experimental value for NH_3 is complicated by the simultaneous occurrence of vibrational relaxation showing almost the same relaxation time. Saksena *et al.*[384] have proposed an alternative treatment of the experimental measurements, giving a distribution of relaxation times. This gives a better fit to the experimental absorption curve, and an estimated value $Z_R = 5{\cdot}1 \pm 2{\cdot}4$. CH_4 ($B = 5{\cdot}25$ cm^{-1}) shows fair agreement with classical prediction at 300 K, and a consistent rise in Z_R as the temperature increases between 77·1 and 300 K. CD_4 ($B = 2{\cdot}64$ cm^{-1}) shows a slightly lower value of Z_R at 300 K in accordance with classical prediction. C_2H_4, with the larger rotation constant, $B = 3{\cdot}98$ cm^{-1}, shows a much lower value of Z_R than CD_4: this is in accord with classical prediction, and may be attributed to the cylindrical shape of the molecule (as contrasted with the spherical-top symmetry of CD_4) leading to a more asymmetric potential and more efficient rotational transition.

The more complex hydrocarbon molecules have very small rotational constants and show low values of Z_R in accord with classical prediction. It will be noted that the classical theories consistently predict values of Z_R which are somewhat larger than the experimental: this may be due to underestimating the 'roughness' of the molecules, or the asymmetry of the potential. The two linear molecules CO_2 and N_2O have low rotational constants, and show the expected small values of Z_R.

5.3. Relaxation in mixtures

Extensive measurements have been made of rotational relaxation of diatomic and simple polyatomic gases in mixtures with different inert gases: these are listed in Table 5.4. The results for p-H_2 at temperatures below 170 K, where the $J = 0{\to}2$ transition predominates, all show the same pattern of Z_R values with a *minimum* for mixtures with Ne: Z_R (H_2) > (He) > (Ne) < (Ar) < (Xe). At 300 K the J=2→4 transition plays a significant role, and Valley and Amme[362] have observed two distinct relaxation times corresponding to collision numbers Z_{20} and Z_{42}. The Z_{20} values show the same pattern as at lower temperatures, with a minimum value $Z_{20} = 74$ for Ne, and there is fair accord with the quantum-mechanical calculations of Agrawal and Saksena.[377,380] The Z_{42} values show a different pattern, with He collisions less efficient than self-collisions: Z_R (H_2) < (He) > (Ne) < (Ar), and this is also in agreement with the quantum-mechanical calculations. Valley and Amme suggest that the comparative inefficiency of He as a collision partner in promoting the $J = 2{\to}4$ transition may be due to a simultaneous transition

$$p\text{-}H_2(J{=}2) + p\text{-}H_2(J{=}2) \to p\text{-}H_2(J{=}0) + p\text{-}H_2(J{=}4) - 473 \text{ cm}^{-1}, \qquad (5.4)$$

enhancing the effect of self-collisions. This would be in accord with the quantum-mechanical prediction of Crawford,[390] discussed in § 5.2.1, but does not seem to be included in the calculations of Agrawal and Saksena.

The measurements on o-D_2 by Jonkman *et al.*[346] are at lower temperatures where the J=0→2 transition will predominate. All inert gas collisions are more efficient than self-collisions, and the efficiency at 77·3 K increases steadily with increasing mass, so that there is no longer a minimum value of Z_R for Ne. Classical theory is not strictly applicable here, but it is interesting to note that eqn (5.2) would predict that D_2, having a larger moment of inertia than H_2,

TABLE 5.4.

Rotational relaxation in mixtures

Mixture A	Mixture B (mol.wt)	Temp. (K)	Z_R (exp.)	Ref.	Z_R (calc.)	Method*	Ref.
p-H_2	p-H_2 (2)	90·5	635	344	–		
	He (4)	90·5	369	344	294	DW	380
	Ne (20)	90·5	**235**	344	281	DW	380
	Ar (40)	90·5	255	344	338	DW	380
p-H_2	p-H_2 (2)	115	604	344	–		
	Kr (80)	115	277	344	–		
p-H_2	p-H_2 (2)	170	411	344	–		
	He (4)	170	207	344	203	DW	380
	Ne (20)	170	**140**	344	207	DW	380
	Ar (40)	170	198	344	301	DW	380
	Kr (80)	170	253	344	–		
	Xe (130)	170	356	344	–		
			(Z_{20}) (Z_{42})		(Z_{20}) (Z_{42})		
p-H_2	p-H_2 (2)	300	220 660	362	329 –	DW	372
	He (4)	300	96 973	362	147 867	DW	377
	Ne (20)	300	**74** 673	362	**96 584**	DW	377
	Ar (40)	300	131 870	362	150 1315	DW	377
o-D_2	o-D_2 (4)	44·5	413	346	–		
	He (4)	44·5	238	346	234	DW	380
o-D_2	o-D_2 (4)	77·3	308	346	–		
	He (4)	77·3	175	346	200	DW	380
	Ne (20)	77·3	96	346	175	DW	380
	Ar (40)	77·3	79	346	181	DW	380
o-D_2	o-D_2 (4)	90·5	288	346	–		
	He (4)	90·5	152	346	175	DW	380
	Ne (20)	90·5	91	346	150	DW	380
	Ar (40)	90·5	95	346	166	DW	380
					(Z_{20}) (Z_{42})		
o-D_2	He (4)	300	–		62 255	DW	377
	Ne (20)	300	–		39 **168**	DW	377
	Ar (40)	300	–		52 324	DW	377
HD	HD (3)	20 to 40	10·3	347	–		
	He (4)	27·0	17·3	347	–		
	He (4)	32·1	16·3	347	–		
	He (4)	39·9	18·1	347	–		
N_2	N_2 (28)	300	4·7	354	–		
	He (4)	300	6·0	354	1·7	Cl (W)	354
	He (4)	308	3·8 to 8·6	361	3·6	Tr	361
	Ne (20)	308	**1·9 to 3·1**	361	**1·9**	Tr	361
	Ar (40)	308	2·5 to 3·6	361	2·3	Tr	361
	Xe (130)	308	3·8 to 5·8	361	4·0	Tr	361

Mixture A	B (mol.wt.)	Temp. (K)	Z_R (exp.)	Ref.	Z_R (calc.)	Method*	Ref.
O_2	O_2 (32)	300	4·1	354	–		
	He (4)	300	7·5	354	1·4	Cl (W)	354
NO	NO (30)	300	~1·0	144	–		
	He (4)	300	~1·0	349	–		
CH_4	CH_4 (16)	308	12·3	363	18	Cl (WCU)	365
	He (4)	308	3·0	363	6	Cl (W)	365
	Ne (20)	308	6·3	363	15	Cl (W)	365
	Ar (40)	308	12·0	363	22	Cl (W)	365
	Xe (130)	308	27·0	363	33	Cl (W)	365
CD_4	CD_4 (20)	308	10·0	364	12	Cl (WCU)	365
	He (4)	308	2·6	364	3	Cl (W)	365
	Ne (20)	308	4·1	364	9	Cl (W)	365
	Ar (40)	308	7·3	364	13	Cl (W)	365
	Xe (130)	308	14·0	364	17	Cl (W)	365

*DW = quantum-mechanical distorted wave calculations
Cl = classical calculations: (W) = Widom, eqn. (5.2), (WCU) = Wang Chang & Uhlenbeck, eqn. (5.3)
Tr = classical trajectory calculations

would show minimum Z_R at a *larger* reduced collision mass. The measurements at 90·5 K show a barely significant difference between Ne and Ar, which accords with the view that the minimum has shifted to the heavier collision partner. The quantum-mechanical calculations for Z_{20} and Z_{42} at 300 K again predict a minimum for collisions with Ne, as they did for p-H_2, but there are no experimental data for comparison.

The measurements on mixtures of HD with He were all made in the temperature range 20 to 40 K, where Z_R for HD has the constant value ~ 10, which was attributed in § 5.2.1 to the predominating influence of the attractive potential. Collisions with He are consistently less efficient, corresponding to a weaker attractive potential.

For N_2, collisions with He are significantly less efficient than self-collisions: this is not in accord with the classical prediction, and suggests that the simultaneous R-R transitions postulated for p-H_2 also play a role in the self-collisions. The values of Z_R for collision with the heavier inert gases are in good agreement with classical trajectory calculations, showing a pronounced maximum efficiency for Ne.[361] For O_2, collisions with He are also less efficient than self-collisions. For CH_4 and CD_4 collisions with He are more efficient than self-collisions, in agreement with classical prediction, and the values of Z_R show fair agreement with calculations based on eqn (5.2).

Fluorescence studies on the rotational relaxation of heavier diatomic molecules in mixtures show conclusive evidence for the occurrence of multiple quantum transitions. Broida and Carrington[385] showed that rotational relaxation of NO ($A^2\ \Sigma^+$) in collision with He, Ar, H_2, N_2, NO, and CO_2 involves multiple transitions at least up to $\Delta J = \pm 5$. Investigation of resonance fluorescence of the heavier I_2 ($B^3\ \Pi_{ou}$) by Steinfeld and Klemperer[386] showed rotational transitions involving $\Delta J \leqslant 40$ to occur from the $J = 34$ level in collision with He, Ne, Ar, Kr, Xe, H_2, O_2, CO_2, CH_3Cl, SO_2, and NH_3.

5.4. The Polanyi-Woodall equation. Resonant R-R transfer

Polanyi and Woodall[67] made detailed spectroscopic observations of the way in which the rotational population changes with time in rotationally excited HCl, produced in the chemiluminescent reaction

$$H + Cl_2 \rightarrow HCl^* + Cl.$$

The reaction produces rotationally excited HCl* with a population peaking at $J = 13$. The room-temperature Boltzmann distribution peaks at $J = 2$ to 3, and under suitable conditions the rate of population change of individual rotational levels during relaxation to Boltzmann equilibrium may be followed spectroscopically. The relaxation is due primarily to HCl* + H_2 collisions (the H atoms being produced by passing a stream of H_2 through a Wood's discharge tube). The observed pattern of relaxation may be accounted for if deactivation is assumed to occur predominantly by small ΔJ changes, and the probability of R-T transfer decreases with increasing J. The only quantitative model which fitted the observations is one in which ΔJ is unrestricted, except for the relation (given earlier in § 3.6)

$$P_{J-J'} = N \exp(-C\Delta E), \qquad (3.17)$$

where $P_{J-J'}$ is the probability of transition from a higher state J to a lower state J', and ΔE is the energy difference between the two states. N and C are constants. No selection rule operates for ΔJ, and transitions involving $\Delta J \leqslant 5$ were observed, but eqn (3.17) predicts that $\Delta J = \pm 1$ transitions will always be the more probable.

Polanyi and Woodall's experiments were conducted at one temperature only, 230 K, and gave a quantitative value $C = 8{\cdot}5 \times 10^{-3}$ cm, whose relation to the Lambert-Salter plot for vibrational relaxation and to the information theoretic approach to energy transfer has been discussed in § 3.6. More recently Ding and Polanyi[392] have used molecular beam technique to investigate rotational transitions at varying collision energies. A supersonic primary beam of HCl or HF was allowed to collide with a target gas. Infrared emission from the beam was recorded, both prior to collision and subsequent to collision, so that the population change in J states could be followed. Collision energies were varied between 1 and 40 kcal mol^{-1}, and the initial rotational population distribution in the beam extended from J=0 to J=15, peaking at J=3. The target gases were He, Ar, Kr, HF, HCl, HBr, HI, H_2S and C_3H_8. The broad picture emerges that a net transfer into low-J states out of high-J states takes place at low collision energies, and the converse at high collision energies, with a significant proportion of transitions having $\Delta J > 1$. The observations can be interpreted quantitatively in terms of eqn (3.17), and give values of C which decrease progressively with increasing collision energy. This is in accord with the information theoretic prediction and analogous to the temperature-dependence for vibrational transitions which follow the Lambert-Salter plot (§ 3.6). At high collision energies, 30-40 kcal mol^{-1}, rotational excitation occurs with a probability comparable to that of elastic scattering.

The systems HCl* + HF and HF* + HCl were both anomalous in showing marked rotational *deactivation* even at the highest collision energies. This may be attributed to R-R transfer, a view which is supported by an alternation observed in the rates of excitation and de-excitation of odd and even J-levels of HCl* in collision with HF. The spectroscopic rotational constant of HCl, $B^* = 10{\cdot}59$ cm^{-1}, is almost exactly half that of HF, $B' = 20{\cdot}94$ cm^{-1}. The rotational levels of a diatomic molecule are given by the relation $E = BJ(J+1)$. If we assume that, at lower J-levels, B' for HF may be taken as exactly double B^* for HCl, the energy levels for the two molecules will be:

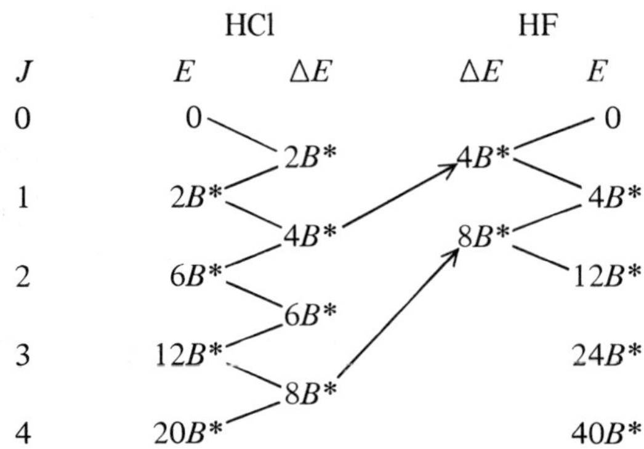

HF as 'target gas' at room temperature will be in a rotational distribution peaking at J=2. There will then be the possibility of resonant R-R transfer from the even rotational levels of HCl*, J=4→3 and J=2→1, but not from the odd levels, J=3→2 or J=1→0. When HCl is the target gas, no such alternation will be observed as *all* rotational transitions for HF* are resonant with a transition in HCl. No similar effect is observed in collisions with either HBr (B=8·47 cm^{-1}) or HI (B=6·55 cm^{-1}), which show no such relation between energy levels. This experiment provides the first conclusive evidence for the efficiency of resonant R-R transfer in systems where appropriate energy levels are present.

5.5. Mechanisms of rotational excitation

The role of rotational excitation in the overall heating and cooling processes of the heavier diatomic and polyatomic gases is uninteresting. R-T transfer is very rapid, multiple quantum transitions occur easily, and the rotational population distribution reaches equilibrium almost as quickly as the translational. For H_2 and hydrides the picture is different; R-T transfer is much slower, also it plays a more important role in the overall heating and cooling processes, as vibrational modes of such molecules do not become active until relatively high temperatures. The rotational excitation process differs fundamentally from the vibrational because of the different ways in which the quantum levels are distributed: the energy gaps between successive vibrational levels become gradually smaller (owing to anharmonicity) as v increases, while the gaps between successive rotational levels become rapidly larger as J increases (cf. Fig. 3.13). This means that R-T transitions, $\Delta J = \pm 1$, become progressively less efficient as J and ΔE increase, while R-R transitions such as (5.4), even when involving neighbouring levels, are nearly always non-resonant and inefficient. The kind of mechanism shown in Fig. 4.21 for the vibrational excitation of cold gases—slow initial activation by V-T transfer to the v=1 level, followed by rapid 'staircase' ascent by near-resonant V-V transfer—does not operate in rotational excitation. Instead, the initial R-T excitation of a cold gas from a low J-level is rapid, but becomes progressively slower as J increases. Conversely, the initial rotational de-excitation of a hot gas, with rotational population peaking in a high J-level, will be very slow, but will become progressively faster as the rotational temperature drops. Very high rotational levels, sometimes produced in chemiluminescent processes, may thus show a high degree of metastability. The rotational temperature of a hot gas, initially in Boltzmann equilibrium, cooling in a jet expansion can lag behind both translational and vibrational temperatures, giving rise to a non-equilibrium distribution: $T_{\text{ROT}} > T_{\text{VIB}} > T_{\text{TR}}$. In contrast, the temperature distribution in a cold gas, undergoing rapid heating in a shock-front, where rotational excitation will be much faster than vibrational, will be $T_{\text{TR}} > T_{\text{ROT}} > T_{\text{VIB}}$. A detailed theoretical study of heating and cooling processes in H_2 in the range $100\ \text{K} \leqslant T \leqslant 1100\ \text{K}$, where only rotational and translational energy are involved, has been made by Rabitz and Lam.[391] Ding and Polanyi[392] have followed spectroscopically the rotational distribution of HF and HCl, initially in rotational levels $J \leqslant 15$, in a supersonic beam expansion, and find clear experimental evidence that the lower J-levels relax much faster than the upper J-levels.

It is interesting to note, in conclusion, that rotational de-excitation by emission of *radiation* follows an entirely different pattern. This can be important in interstellar clouds, where low temperature and extremely small molecular concentrations can render collisional relaxation extremely slow. The probability for radiative transition between the J' and $(J'-1)$ levels varies with J^3 (cf. § 2.4.1). The upper J-levels are therefore depopulated much more rapidly than the lower levels, and a population inversion builds up between the J=1 and J=0 levels.[187] This can lead to maser action.

REFERENCES

1 Hirschfelder, J.O., Curtiss, C.F., & Bird, R.B. (1954). *Molecular theory of gases and liquids*, Wiley, New York
2 Rowlinson, J.S. (1961). *Molec. Phys.* **4**, 317
3 Ehrenfest, P. (1914). *Koninkl. Ned. Akad. Wetenschap. Proc.* **16**, 591
4 Landau, L. & Teller, E. (1936). *Phys. Z. Sowj. Un.* **10**, 34
5 Herzfeld, K.F. & Litovitz, T.A. (1959). *Absorption and dispersion of ultrasonic waves*, Academic Press, New York
6 Rowlinson, J.S. (1949). *Trans. Faraday Soc.* **45**, 980
7 Lambert, J.D., Roberts, G.A.H., Rowlinson, J.S., & Wilkinson, V.J. (1949). *Proc. R. Soc.* **A196**, 113
8 Taylor, H.S. & Glasstone, S. (1951). *Treatise on physical chemistry*, Van Nostrand, New York, Vol. 2, p.339
9 Pauling, L. (1940). *Nature of the chemical bond*, Cornell University Press, Chap. 10.
10 Inskeep, R.G., Kelliker, J.M., McMahon, P.E., & Somers, B.G. (1958). *J. chem. Phys.* **28**, 1033
11 Dinerman, C.E. & Ewing, G.E. (1970). *J. chem. Phys.* **53**, 626
12 Singh, Y., Deb, S.K., & Barua, A.K. (1967). *J. chem. Phys.* **46**, 4036
13 Billingsley, J. & Callear, A.B. (1971). *Trans. Faraday Soc.* **67**, 257
14 MacCormack, K.E. & Schneider, W.G. (1951). *J. chem. Phys.* **19**, 849
15 Pople, J.A. (1954). *Proc. R. Soc.* **A221**, 508
16 Stogryn, D.E. & Hirschfelder, J.O. (1959). *J. chem. Phys.* **31**, 1531
17 Leckenby, R.E. & Robbins, E.J. (1966). *Proc. R. Soc.* **A291**, 389
18 Kudian, A., Welsh, H.S., & Watanabe, A. (1965). *J. chem. Phys.* **43**, 3397
19 Pal, A.K. & Barua, A.K. (1967). *Trans. Faraday Soc.* **63**, 341
20 Cottrell, T.L. & McCoubrey, J.C. (1961). *Molecular energy transfer in gases*, Butterworths, London
21 Parker, J.G. (1959). *Physics Fluids* **2**, 449
22 Mason, E.A. & Monchick, L. (1962). *J. chem. Phys.* **36**, 1622
23 Stretton, J.L. (1969). *Transfer and storage of energy by molecules*, ed. G.M. Burnett & A.M. North, Wiley-Interscience, Vol. 2, chaps. 1 and 2
24 Roberts, J.K. (1940). *Heat and thermodynamics*, Blackie, Chap. 6
25 Dymond, J.H. & Smith, E.B. (1969). *The virial coefficients of gases*, Clarendon Press, Oxford
26 Greenspan, M. (1965). *Physical acoustics*, ed. W.P. Mason, Academic Press, New York, chap. 1 and refs.
27 Tanczos, F.L. (1956). *J. chem. Phys.* **25**, 439
28 Dickens, P.G. & Linnett, J.W. (1957). *Proc. R. Soc.* **A243**, 84
29 Jones, D.G. (1967). D. Phil. Thesis (Oxford)
30 Lambert, J.D. & Rowlinson, J.S. (1950). *Proc. R. Soc.* **A204**, 424
31 Martinez, J.V., Strauch, J.G., & Decius, J.C. (1964). *J. chem. Phys.* **40**, 186

32 Strauch, J.G. & Decius, J.C. (1966). *J. chem. Phys.* **44**, 339
33 Hancock, J.K. & Decius, J.C. (1969). *J. chem. Phys.* **51**, 5374
34 Edmonds, P.D. & Lamb, J. (1958). *Proc. phys. Soc.* **71**, 17
35 Edmonds, P.D. & Lamb, J. (1958). *Proc. phys. Soc.* **72**, 940
36 Holmes, R., Smith, F.A., & Tempest, W. (1963). *Proc. phys. Soc.* **81**, 311
37 Parker, J.G. (1961). *J. chem. Phys.* **34**, 1763
38 Lukasik, S.J. & Young, J.E. (1957). *J. chem. Phys.* **27**, 1149
39 (a) Bradley, J.N. (1962). *Shock waves in chemistry and physics*, Methuen, London
(b) Gaydon, A.G. & Hurle, I.R. (1963). *The shock tube in high temperature chemistry and physics*, Chapman & Hall, London
40 Greene, E.F. & Toennies, J.P. (1964). *Chemical reactions in shock waves*, Arnold, London
41 Borell, P. (1969). *Transfer and storage of energy by molecules*, ed. G.M. Burnett & A.M. North, Wiley-Interscience, New York, chap. 3
42 Miller, R.C. & Nordland, W.A. (1967). *Appl. Phys. Lett.* **10**, 53
43 Wallace, R.W. (1970). *Appl. Phys. Lett.* **17**, 497
44 Ambartzumian, R.V., Letokhov, V.S., Makarov, G.N., & Chekalin, N.V. (1972). *Chem. Phys. Lett.* **13**, 49
45 Moore, C.B. (1967). *Fluorescence*, ed. C.G. Guilbault, Marcel Dekker, New York, chap. 3
46 Moore, C.B. (1973). *Adv. chem. Phys.* **23**, 41
47 Millikan, R.C. (1963). *J. chem. Phys.* **38**, 2855
48 Hooker, W.J. & Millikan, R.C. (1963). *J. chem. Phys.* **38**, 214
49 Steinfeld, J.I., Burak, I., Sutton, D.G., & Novak, A.V. (1970). *J. chem. Phys.* **52**, 5421
50 Oka, T. (1967). *J. chem. Phys.* **47**, 13
51 Frey, R., Jukasik, J., & Ducuing, J. (1972). *Chem. Phys. Lett.* **14**, 514
52 De Martini, F. & Ducuing, J. (1966). *Phys. Rev. Lett.* **17**, 117
53 Tyndall, J. (1881). *Proc. R. Soc.* **31**, 307, 408
54 Slobodskaya, P.V. (1948). *Izv. Akad. Nauk. SSSR, Ser. Fiz.* **12**, 656
55 Read, A.W. (1968). *Adv. Mol. Relax. Processes* **1**, 257
56 Burak, I., Houston, P., Sutton, D.G., & Steinfeld, J.I. (1970). *J. chem. Phys.* **53**, 3632
57 Huetz-Aubert, M. & Tripodi, R. (1971). *J. chem. Phys.* **55**, 5724
58 Cottrell, T.L., Macfarlane, I.M., Read, A.W., & Young, A.H. (1966). *Trans. Faraday Soc.* **62**, 266
59 Shimizu, T. & Oka, T. (1970). *Phys. Rev.* **A2**, 1177
60 Fink, E.H., Akins, D.L., & Moore, C.B. (1972). *J. chem. Phys.* **56**, 900
61 Basco, N., Callear, A.B., & Norrish, R.G.W. (1961). *Proc. R. Soc.* **A260**, 459
62 Donovan, R.J. & Husain, D. (1967). *Trans. Faraday Soc.* **63**, 2879
63 Callear, A.B. & Williams, G.J. (1966). *Trans. Faraday Soc.* **62**, 2030
64 Donovan, R.J., Husain, D., & Stevenson, C.D. (1970). *Trans. Faraday Soc.* **66**, 2148
65 Anlauf, K.G., Kuntz, P.J., Maylotte, D.H., Pacey, P.D., & Polanyi, J.C. (1967). *Disc. Faraday Soc.* **44**, 183
66 Hancock, G. & Smith, I.W.M. (1971). *Appl. Optics* **10**, 1827
67 Polanyi, J.C. & Woodall, K.B. (1972). *J. chem. Phys.* **56**, 1563
68 Corran, P.G., Lambert, J.D., Salter, R., & Warburton, B. (1958). *Proc. R. Soc.* **A244**, 212
69 Shin, H.K. (1968). *J. Am. chem. Soc.* **90**, 3029

70 Toennies, J.P. (1974). *Chem. Soc. Rev.* **3**, 407
71 Mariella, R.P., Herschbach, D.R., & Klemperer, W. (1974). *J. chem. Phys.* **61**, 4575
72 Toennies, J.P. (1965). *Z. Phys.* **182**, 257
73 Ding, A.M.G. & Polanyi, J.C. (1973). *Faraday Disc. Chem. Soc.* **55**, 225
74 Scott, P.B., Mincer, T.R., & Muntz, E.P. (1973). *Chem. Phys. Lett.* **22**, 71
75 Chapman, S. & Cowling, T.G. (1939). *Mathematical theory of non-uniform gases*, Cambridge Univ. Press
76 Eucken, A. (1913). *Phys. Z.* **14**, 324
77 Cowling, T.G. (1964). *Brit. J. appl. Phys.* **15**, 959
78 Jackson, J.M. & Mott, N.F. (1932). *Proc. R. Soc.* **A137**, 703
79 Takayanagi, K. (1952). *Progr. theor. Phys., Kyoto* **8**, 497
80 Schwartz, R.N., Slawsky, Z.I., & Herzfeld, K.F. (1952). *J. chem. Phys.* **20**, 1591
81 Schwartz, R.N. & Herzfeld, K.F. (1954). *J. chem. Phys.* **22**, 767
82 Wilson, E.B., Decius, J.C., & Cross, P.C. (1955). *Molecular vibrations* McGraw Hill, New York
83 Shin, H.K. (1968). *J. chem. Phys.* **48**, 3644
84 Takayanagi, K. (1965). *Adv. Atom. Molec. Phys.* **1**, 149
85 Rapp, D. & Kassal, T. (1969). *Chem. Rev.* **69**, 61
86 Secrest, D. (1973). *A. Rev. phys. Chem.* **24**, 379
87 Parks-Smith, D.G. (1964). D. Phil. Thesis, Oxford
88 Stretton, J.L. (1964). D. Phil. Thesis, Oxford
89 Herzberg G. (1945). *Infrared and Raman spectra of polyatomic molecules*, Van Nostrand, New York
90 Calvert, J.B. & Amme, R.C. (1966). *J. chem. Phys.* **45**, 4710
91 Shin, H.K. (1972). *J. chem. Phys.* **57**, 1363
92 Millikan, R.C. & White, D.R. (1963). *J. chem. Phys.* **39**, 98
93 (a) Blackman, V.H. (1956). *J. Fluid Mech.* **1**, 61
(b) Gaydon, A.G. & Hurle, I.R. (1962). *8th Symposium on Combustion*, Williams & Wilkins, Baltimore
(c) Clouston, J.G., Gaydon, A.G., & Glass, I.I. (1958). *Proc. R. Soc.* **A248**, 429
94 Lukasik, S.J. & Young, J.E. (1957). *J. chem. Phys.* **27**, 1149
95 Henderson, M.C. (1962). *J. acoust. Soc. Amer.* **34**, 349
96 Hooker, W.J. & Millikan, R.C. (1963). *J. chem. Phys.* **38**, 214
97 Matthews, D.L. (1961). *J. chem. Phys.* **34**, 639
98 Millikan, R.C. (1963). *J. chem. Phys.* **38**, 2855
99 Ferguson, M.G. & Read, A.W. (1965). *Trans. Faraday Soc.* **61**, 1559
100 Generalov, N.A. (1963). *Dokl. Akad. Nauk. SSSR.* **148**, 373
101 Parker, J.G. (1964). *J. chem. Phys.* **41**, 1600
102 Camac, M. (1961). *J. chem. Phys.* **34**, 448
103 White, D.R. & Millikan, R.C. (1963). *J. chem. Phys.* **39**, 1803
104 Shields, F.D. & Lee, K.P. (1963). *J. acoust. Soc. Amer.* **35**, 251
105 Roesler, H. (1966). *Acustica* **17**, 73
106 Holmes, R., Smith, F.A., & Tempest, W. (1963). *Proc. Phys. Soc.* **81**, 311
107 Parker, J.G. (1961). *J. chem. Phys.* **34**, 1763
108 Smiley, E.F. & Winkler, E.H. (1954). *J. chem. Phys.* **22**, 2018
109 Shields, F.D. (1960). *J. acoust. Soc. Amer.* **32**, 180
110 Eucken, A. & Becker, R. (1934). *Z. phys. Chem.* **B27**, 235
111 Sittig, E. (1960). *Acustica* **10**, 81

112 Richardson, E.G. (1959). *J. acoust. Soc. Amer.* **31**, 152
113 Dickens, P.G. & Ripamonti, A. (1961). *Trans. Faraday Soc.* **57**, 735
114 Kiefer, J.H. & Lutz, R.W. (1966). *J. chem. Phys.* **44**, 669
115 Dove, J.E. & Teitelbaum, H. (1974). *Chem. Phys.* **6**, 431
116 Audibert, M.M., Joffrin, C., & Ducuing, J. (1974). *Chem. Phys. Lett.* **25**, 158
117 Calvert, J.B. (1972). *J. chem. Phys.* **56**, 5071
118 Lukasik, J. & Ducuing, J. (1974). *Chem. Phys. Lett.* **27**, 203
119 Simpson, C.J.S.M., Price, T.J., & Crowther, M.E. (1975). *Chem. Phys. Lett.* **34**, 181
120 Audibert, M.M., Vilaseca, R., Lukasik, J., & Ducuing, J. (1975). *Chem. Phys. Lett.* **31**, 232
121 Sorensen, G.B. (1974). *Chem. Phys.* **5**, 244
122 Bott, J.F. & Cohen, J. (1971). *J. chem. Phys.* **55**, 3698
123 Bott, J.F. (1972). *J. chem. Phys.* **57**, 96
124 Lucht, R.A. & Cool, T.A. (1974). *J. chem. Phys.* **60**, 1026
125 Stephens, R.R. & Cool, T.A. (1972). *J. chem. Phys.* **56**, 5863
126 Hinchen, J.J. (1973). *J. chem. Phys.* **59**, 233
127 Ahl, J.L. & Cool, T.A. (1973). *J. chem. Phys.* **58**, 5540
128 Breshears, W.D. & Bird, R.F. (1969). *J. chem. Phys.* **50**, 333
129 Zittel, B.M. & Moore, C.B. (1973). *J. chem. Phys.* **59**, 6636
130 Kiefer, J.H., Breshears, W.D., & Bird, P.F. (1969). *J. chem. Phys.* **50**, 3641
131 Cottrell, T.L., Dobbie, R.C., McLain, J., & Read, A.W. (1964). *Trans. Faraday Soc.* **60**, 241
132 Moore, C.B. (1965). *J. chem. Phys.* **43**, 2979
133 Shin, H.K. (1971). *J. phys. Chem.* **76**, 1079
134 Bott, J.F. & Cohen, N. (1973). *J. chem. Phys.* **58**, 934
135 Smith, D.F. (1959). *J. Molec. Spectrosc.* **3**, 473
136 Rank, D.H., Sitaram, P., Glickman, W.A., & Wiggins, T.A. (1963). *J. chem. Phys.* **39**, 2673
137 Breshears, W.D. & Bird, P.F. (1970). *J. chem. Phys.* **52**, 999
138 Shin, H.K. (1973). *J. chem. Phys.* **59**, 879
139 Dinerman, C.F. & Ewing, G.E. (1970). *J. chem. Phys.* **53**, 626
140 Williams, J.E. & Murrell, J.N. (1971). *J. Am. chem. Soc.* **93**, 7149
141 Wray, K.L. (1962). *J. chem. Phys.* **36**, 2597
142 Bradley, J.N. & Lewis, D. (1969). *J. chem. Phys.* **50**, 544
143 Amme, R.L., Fanchi, J.R., & Olson, J.R. (1973). *J. chem. Phys.* **58**, 4707
144 Bauer, H.-J. & Sahm, K.F. (1965). *J. chem. Phys.* **42**, 3400
145 Billingsley, J. & Callear, A.B. (1971). *Trans. Faraday Soc.* **67**, 257
146 Nikitin, E.G. (1960). *Optics Spectrosc.* **9**, 8
147 Eucken, A. & Aybar, S. (1940). *Z. phys. Chem.* **B46**, 195
148 Magigorsky, W.M. (1963). *J. chem. Phys.* **39**, 2704
149 Richards, L.W. & Sigafoos, D.H. (1965). *J. chem. Phys.* **43**, 492
150 Hill, G.L. & Winter, T.G. (1968). *J. chem. Phys.* **49**, 440
151 Cottrell, T.L. & Matheson, A.J. (1962). *Trans. Faraday Soc.* **58**, 2336
152 Stretton, J.L. (1965). *Trans. Faraday Soc.* **61**, 1053
153 Schäfer, K. (1940). *Z. phys. Chem.* **B46**, 212
154 Simpson, C.J.S.M., Bridgman, K.B., & Chandler, T.R.D. (1968). *J. chem. Phys.* **49**, 513
155 Fogg, P.G.T., Hanks, P.A., & Lambert, J.D. (1953). *Proc. R. Soc.* **A219**, 490
156 Griffith, W., Brickl, D., & Blackman, V. (1956). *Phys. Rev.* **102**, 1209

157 Eucken, A. & Nümann, E. (1937). *Z. phys. Chem.* **B36**, 163
158 Buschmann, K.F. & Schäfer, K. (1941). *Z. phys. Chem.* **B50**, 73
159 Lambert, J.D. & Salter, R. (1957). *Proc. R. Soc.* **A243**, 78
160 Bass, H.E., Olson, J.R., & Amme, R.C. (1974). *J. acoust Soc. Amer.* **56**, 1455
161 Roesler, H. & Sahm, K.F. (1965). *J. acoust. Soc. Amer.* **37**, 386
162 Fujii, Y., Lindsay, R.B., & Urushihara, J. (1963). *J. acoust. Soc. Amer.* **35**, 961
163 Bass, H.E. & Winter, T.G. (1972). *J. chem. Phys.* **56**, 3619
164 Edwards, A.J. (1959). B.Sc. Thesis, Oxford, cf. ref. 20
165 Lambert, J.D. & Salter, R. (1959). *Proc. R. Soc.* **A253**, 277
166 Lambert, J.D. (1972). *J. chem. Soc. Faraday II* **68**, 364
167 Cottrell, T.L. & Matheson, A.J. (1963). *Trans. Faraday Soc.* **59**, 824
168 Angona, F.A. (1953). *J. acoust. Soc. Amer.* **25**, 1116
169 Rossing, T.D. & Legvold, S. (1955). *J. chem. Phys.* **23**, 1118
170 Hanks, P.A. (1951). D.Phil. Thesis, Oxford, cf. ref. 155
171 Fogg, P.G.T. (1954). D. Phil. Thesis, Oxford, cf. Fogg, P.G.T. & Lambert J.D. (1955). *Proc. R. Soc.* **A232**, 537
172 Seshagiri Rao, T.S. & Srinivasachari, E. (1968). *Ind. J. Pure Appl. Phys.* **6**, 322
173 Seshagiri Rao, T. & Srinivasachari, E. (1966). *Ibid.* **4**, 430
174 Holmes, R., Jones, G.R., & Pusat, N. (1964). *J. chem. Phys.* **41**, 2512
175 Sette, D., Busala, A., & Hubbard, J.C. (1955). *J. chem. Phys.* **23**, 787
176 Seshagiri Rao, T. & Hubbard, J.C. (1955). *J. acoust. Soc. Amer.* **27**, 231
177 Jones, D.G., Lambert, J.D., Saksena, M.P., & Stretton, J.L. (1969). *Trans. Faraday Soc.* **65**, 965
178 Holmes, R., Jones, G.R., & Lawrence, R. (1966). *Trans. Faraday Soc.* **62**, 46
179 Hudson, G.H., McCoubrey, J.C., & Ubbelohde, A.R. (1961). *Proc. R. Soc.* **A264**, 289
180 Merill, K.M. & Amme, R.C. (1969). *J. chem. Phys.* **51**, 844
181 Arnold, J.W., McCoubrey, J.C., & Ubbelohde, A.R. (1958). *Proc. R. Soc.* **A248**, 445
182 Lambert, J.D., Parks-Smith, D.G. & Stretton J.L. (1964). *Proc. R. Soc.* **A282**, 380
183 Lambert, J.D., Parks-Smith, D.G., & Stretton, J.L. (1970). *Trans. Faraday Soc.* **66**, 2720
184 Hinsch, H. (1961). *Acustica* **11**, 230
185 Seshagiri Rao, T. & Srinivasachari, E. (1965). *Ind. J. Pure Appl. Phys.* **3**, 296
186 Haebel, E.U. (1965). *Acustica* **15**, 426
187 Goldsmith, P. (1972). *Astrophys. J.* **176**, 597
188 Hansen, F.C. & Pearson, W.E. (1970). *J. chem. Phys.* **53**, 2557
189 Neilsen, W.B. & Gordon, R.G. (1973). *J. chem. Phys.* **58**, 4131, 4149
190 Ottinger, C.H. & Poppe, D. (1971). *Chem. Phys. Lett.* **8**, 513
191 Winter, T.G. & Bass, H.E. (1970). *J. acoust. Soc. Amer.* **48**, 1119
192 Bauer, H.-J., Paphitis, A.C.C., & Schotter, R. (1970). *Physica* **47**, 109
193 Olson, J.R. & Legvold, S. (1963). *J. chem. Phys.* **39**, 2902
194 Hamann, S.D. & Lambert, J.A. (1954). *Aust. J. Chem.* **7**, 1
195 Kurzel, R.B., Steinfeld, J.I., Hatzenbuhler, D.A., & Leroi, G.E. (1971). *J. chem. Phys.* **55**, 4822
196 Poulsen, L.L., Houston, P.L., & Steinfeld, J.I. (1973). *J. chem. Phys.* **58**, 3381

197 Holmes, R., Smith, F.A., & Tempest, W. (1964). *Proc. phys. Soc.* **83**, 769
198 Lewis, J.W.L. & Shields, F.D. (1967). *J. acoust. Soc. Amer.* **41**, 100
199 Welsh, A.G. & Taylor, J.E. (1970). *J. chem. Phys.* **53**, 2529
200 Eucken, A. & Jaacks, H. (1935). *Z. phys. Chem.* **B30**, 85
201 Millikan, R.C. (1964). *J. chem. Phys.* **40**, 2594
202 Miller, D.J. & Millikan, R.C. (1970). *J. chem. Phys.* **53**, 3384
203 Shields, F.D. & Lee, K.P. (1964). *J. chem. Phys.* **40**, 737
204 Steele, R.V. & Moore, C.B. (1974). *J. chem. Phys.* **60**, 2794
205 Shields, F.D. (1967). *J. chem. Phys.* **46**, 1063
206 Stephenson, J.C. (1973). *J. chem. Phys.* **59**, 1523
207 Kneser, H.O., Bauer, H.-J., & Koschi, H. (1967). *J. acoust. Soc. Amer.* **41**, 1029
208 Kamimoto, G. & Matsui, H. (1970). *J. chem. Phys.* **53**, 3987
209 Houston, P.L., Nowak, A.V., & Steinfeld, J.I. (1973). *J. chem. Phys.* **58**, 3373
210 Eucken, A. & Nümann, E. (1937). *Z. phys. Chem.* **B36**, 163
211 Lewis, J.W.L. & Lee, K.P. (1965). *J. acoust. Soc. Amer.* **38**, 813
212 Eucken, A. & Küchler, L. (1938). *Z. tech. Phys.* **19(12)**, 517
213 Widom, B. & Bauer, S.H. (1953). *J. chem. Phys.* **21**, 1670
214 Hancock, J.K. & Green, W.H. (1972). *J. chem. Phys.* **57**, 4515
215 Breshears, W.D. & Bird, P.F. (1968). *J. chem. Phys.* **48**, 4768
216 Eckstrom, D.J. (1973). *J. chem. Phys.* **59**, 2787
217 McNeal, R.J., Whittson, M.E., & Cook, G.R. (1972). *Chem. Phys. Lett.* **16**, 507
218 Kiefer, I.H. & Lutz, R.W. (1968). *11th. Symposium on Combustion*, Berkeley, Calif. p.67
219 Center, R.E. (1973). *J. chem. Phys.* **58**, 5230
220 Center, R.E. (1973). *J. chem. Phys.* **59**, 3523
221 Cramp, J.H.W. & Lambert, J.D. (1973). *Chem. Phys. Lett.* **22**, 146
222 von Rosenberg, C.W. & Wray, K.L. (1971). *J. chem. Phys.* **54**, 1406
223 Blair, L.S., Breshears, W.D., & Schott, G.L. (1973). *J. chem. Phys.* **59**, 1582
224 Quigley, G.P. & Wolga, G.J. (1974). *Chem. Phys. Lett.* **27**, 276
225 Breig, E.L. (1969). *J. chem. Phys.* **51**, 4539
226 Thompson, D.L. (1972). *J. chem. Phys.* **57**, 4164
227 Brillouin, L. (1962). *Science and information theory*, Academic Press, New York, Ch. 12
228 Benson, R.C., Benard, D.J., & Walker, R.E. (1974). *J. chem. Phys.* **61**, 1652
229 Nikitin, E.E. & Umanski, S.Ya. (1972). *Faraday Disc. Chem. Soc.* **53**, 7
230 Thommarsen, R.L. & Berend, G.C. (1974). *Int. J. chem. Kinet.* **6**, 597
231 Udseth, H., Giese, C.F., & Gentry, W.P. (1973). *Phys. Rev.* **A8**, 2483
232 Kung, R.T.V. & Center, R.E. (1975). *J. chem. Phys.* **62**, 2187
233 Buchwald, M.I. & Wolga, G.J. (1975). *J. chem. Phys.* **62**, 2828.
234 Quigley, G.P. & Wolga, G.J. (1975). *J. chem. Phys.* **62**, 4560
235 Heidner, R.F. & Bott, J.F. (1975). *J. chem. Phys.* **63**, 1810
236 Steele, R.V. & Moore, C.B. (1974). *J. chem. Phys.* **60**, 2794
237 Brown, R.D.H., Glass, G.P., & Smith, I.W.M. (1975). *Chem. Phys. Lett.* **32**, 517
238 Macdonald, R.G., Moore, C.B., Smith, I.W.M., & Woodarczyk, F.J. (1975). *J. chem. Phys.* **62**, 2934
239 Millikan, R.C. & White, D.R. (1963). *J. chem. Phys.* **39**, 98
240 Millikan, R.C. & White, D.R. (1963). *J. chem. Phys.* **39**, 1807

241 Moore, C.B., Wood, R.E., Bei-lok, Hu., & Yardley, J.T. (1967). *J. chem. Phys.* **46**, 4222
242 Lang, J.R., George, T.F., Zimmerman, I.H., & Ying-Wei Liu (1975). *J. chem. Phys.* **63**, 842
243 Wilkins, R.L. (1975). *J. Chem. Phys.* **63**, 534
244 Bauer, S.H. & Tsang, S.C. (1963). *Physics Fluids* **6**, 182
245 (a) Levine, R.D. & Bernstein, R.B. (1974). *Acc. chem. Res.* **7**, 393
(b) Levine, R.D. & Bernstein, R.B. (1975). *Modern theoretical chemistry*, ed. W.H. Miller, Plenum, New York. Vol. 3, Dynamics of molecular collisions
246 Rubison, M. & Steinfeld, J.I. (1974). *Chem. Phys.* **4**, 467
247 Procaccia, I. & Levine, R.D. (1975). *Chem. Phys. Lett.* **33**, 5
248 Procaccia, I. & Levine, R.D. (1975). *J. chem. Phys.* **62**, 2496, 3819
249 Osgood, R.M., Sackett, P.B., & Javan, A. (1974). *J. chem. Phys.* **60**, 1464
250 Bernstein, R.B. (1975). *J. chem. Phys.* **62**, 4570
251 Kurzel, R.B. & Steinfeld, J.I. (1970). *J. chem. Phys.* **53**, 3293
252 Thompson, D.L. (1974). *J. chem. Phys.* **60**, 4557
253 Behnen, S.W., Rothwell, H.L., & Amme, R.C. (1971). *Chem. Phys. Lett.* **8**, 318
254 Simpson, C.J.S.M. & Simmie, J.M. (1971). *Proc. R. Soc.* **A325**, 197
255 Millikan, R.C. & Osburg, L.A. (1964). *J. chem. Phys.* **41**, 2196
256 Sharma R.D. & Kern, C.W. (1971). *J. chem. Phys.* **55**, 1171
257 Breshears, W.D. & Bird, P.F. (1969). *J. chem. Phys.* **51**, 3660
258 Shields, F.D. & Burks, J.A. (1968). *J. acoust. Soc. Amer.* **43**, 510
259 Hancock, J.K. & Green, W.H. (1972). *J. chem. Phys.* **56**, 2474
260 Shin, H.K. (1974). *J. chem. Phys.* **60**, 2167
261 Yardley, J.T. & Moore, C.B. (1968). *J. chem. Phys.* **49**, 1111
262 Weitz, E. & Flynn, G.W. (1973). *J. chem. Phys.* **58**, 2781
263 Grabiner, F.R. & Flynn, G.W. (1974). *J. chem. Phys.* **60**, 398
264 Knudtson, J.T. & Flynn G.W. (1972). *J. chem. Phys.* **57**, 4174
265 Rosen, D.I. & Cool, T.A. (1975). *J. chem. Phys.* **62**, 6097
266 Gamss, L.A. & Ronn, A.M. (1975). *Chem. Phys.* **5**, 319
267 Shields, F.D. & Anderson, B. (1971). *J. chem. Phys.* **55**, 2636
268 Monchick, L. & Mason, E.A. (1961). *J. chem. Phys.* **35**, 1676
269 Dickens, P.G. & Linnett, J.W. (1957). *Proc. R. Soc.* **A243**, 84
270 Siebert, D.R. & Flynn, G.W. (1975). *J. chem. Phys.* **62**, 1212
271 Siebert, D.R., Grabiner, F.R., & Flynn, G.W. (1974). *J. chem. Phys.* **60** 1564
272 Dickens, P.G. & Schofield, D. (1961). *J. chem. Phys.* **35**, 374
273 Meyer, N.J. (1960). *J. chem. Phys.* **33**, 487
274 Hageseth, G.T. (1967). *J. acoust. Soc. Amer.* **42**, 844
275 Dickens, P.G. (1959). *Molec. Phys.* **2**, 206
276 Valley, L.M. & Legvold, S. (1962). *J. chem. Phys.* **36**, 481
277 Seshagiri Rao, T. & Srinivasachari, E. (1965). *Nature*, Lond. **206**, 926
278 Lambert, J.D., Edwards, A.J., Pemberton, D., & Stretton, J.L. (1962). *Disc. Faraday Soc.* **33**, 61
279 Jones, D.G., Lambert, J.D., & Stretton, J.L. (1965). *Proc. phys. Soc.* **86**, 857
280 Zittel, P.F. & Moore, C.B. (1973). *J. chem. Phys.* **58**, 2922
281 Callear, A.B. (1965). *Appl. Optics, Suppl. 2*, 145
282 Millikan, R.C. (1965). *J. chem. Phys.* **43**, 1439
283 Chen, H.L. & Moore, C.B. (1971). *J. chem. Phys.* **54**, 4080

284 Chen, H.L. (1971). *J. chem. Phys.* **55**, 5557
285 Stephenson, J.C. & Moore, C.B. (1972). *J. chem. Phys.* **56**, 1295
286 Guegen, H., Arditi, I., Magottin-Maclou, M., Doyenette, L., & Henry, L. (1971). *C.r. hebd. Séanc. Acad. Sci., Paris.* **272B**, 1139
287 Starr, D.G., Hancock, J.K., & Green, W.H. (1974). *J. chem. Phys.* **61**, 5421
288 Tsuchiya, S. (1964). *Bull. chem. Soc. Japan* **37**, 828 (calc. ref. 287)
289 Starr, D.F. & Hancock, J.K. (1975). *J. chem. Phys.* **63**, 4720
290 Starr, D.F. & Hancock, J.K. (1975). *J. chem. Phys.* **62**, 3747
291 Powell, H.T. (1975). *J. chem. Phys.* **63**, 2635
292 Mahan, B.H. (1967). *J. chem. Phys.* **46**, 98
293 Sharma, R.D. (1969). *Phys. Rev.* **177**, 102
294 Jones, D.G., Lambert, J.D., & Stretton, J.L. (1965). *J. chem. Phys.* **43**, 4541
295 Sharma, R.D. (1975). *Chem. Phys. Lett.* **30**, 261
296 Richman, D.C. & Millikan, R.C. (1974). *J. chem. Phys.* **61**, 4263
297 Stretton, J.L. & Parks-Smith, D.G. Unpublished (cf. ref. 23)
298 Callear, A.B. (1962). *Disc. Faraday Soc.* **33**, 28 and 93
299 Procaccia, I. & Levine, R.D. (1975). *J. chem. Phys.* **63**, 4261
300 Airey, J.R. & Smith, I.W.M. (1972). *J. chem. Phys.* **57**, 1669
301 Kwok, M.A. & Cohen, N. (1974). *J. chem. Phys.* **61**, 5221
302 Hopkins, B.M. & Chen, H-L. (1972). *J. chem. Phys.* **57**, 3816
303 Ridley, B.A. & Smith, I.W.M. (1972). *J. chem. Soc. Faraday II* **68**, 1231
304 Chen, H-L. & Chen, M.Y. (1972). *J. chem. Phys.* **56**, 3315
305 Donovan, R.J., Husain, D., & Stevenson, C.D. (1970). *Trans. Faraday Soc.* **66**, 2148
306 Hancock, J.K. & Green, W.H. (1972). *J. chem. Phys.* **56**, 2474
307 Bott, J.F. & Cohen, N. (1973). *J. chem. Phys.* **58**, 4539
308 Hancock, J.K. & Green, W.H. (1973). *J. chem. Phys.* **59**, 6350
309 Dillon, T.A. & Stephenson, J.C. (1973). *J. chem. Phys.* **58**, 2056
310 Morley, C. & Smith, I.W.M. (1971). *Trans. Faraday Soc.* **67**, 2575
311 Taylor, R.L. & Bitterman, S. (1969). *J. chem. Phys.* **50**, 1720
312 Sato, Y. & Tsuchiya, S. (1970). *Chem. Phys. Lett.* **5**, 293
313 Rosser, W.F., Wood, W.D., & Gerry, E.T. (1969). *J. chem. Phys.* **50**, 4996
314 Sharma, R.D. & Brau, C.A. (1969). *J. chem. Phys.* **50**, 924
315 Stephenson, J.C., Wood, R.E., & Moore, C.B. (1968). *J. chem. Phys.* **48**, 4790
316 Stephenson, J.C., Wood, R.E., & Moore, C.B. (1972). *J. chem. Phys.* **56**, 4813
317 Roach, J.F. & Smith, W.R. (1969). *J. chem. Phys.* **50**, 4114
318 Yardley, J.T. (1968). *J. chem. Phys.* **49**, 2816
319 Sharma, R.D. (1970). *Phys. Rev.* **2A**, 173
320 Andreev, E.A. (1971). *Chem. Phys. Lett.* **11**, 429
321 Noter, Y., Burak, I., & Szöke, A. (1973). *J. chem. Phys.* **59**, 970
322 Stephenson, J.C. & Mosburg, E.R. (1974). *J. chem. Phys.* **60**, 3652
323 Fushiki, Y. & Tsuchiya, S. (1973). *Chem. Phys. Lett.* **22**, 47
324 Powell, H.T. (1973). *J. chem. Phys.* **59**, 4937
325 Jeffers, W.Q. & Kelley, J.D. (1971). *J. chem. Phys.* **55**, 4433
326 Smith, I.W.M. & Wittig, C. (1973). *J. chem. Soc. Faraday II* **69**, 939
327 Dillon, T.A. & Stephenson, J.C. (1972). *Phys. Rev.* **A6**, 1460
328 Chow, C.C. & Greene, E.F. (1965). *J. chem. Phys.* **43**, 324
329 Norrish, R.G.W. (1962). *Disc. Faraday Soc.* **33**, 87
330 Montroll, E.W. & Shuler, K.E. (1957). *J. chem. Phys.* **26**, 454

331 Shuler, K.E. (1960). *J. chem. Phys.* **32**, 1692
332 Treanor, C.E., Rich, J.W., & Rehm, R.G. (1968). *J. chem. Phys.* **48**, 1798
333 Bray, K.N.C. (1970). *J. Phys.* **B3**, 1515
334 Rich, J.W. (1971). *J. appl. Phys.* **42**, 2719
335 Legay, F., Legay-Sommaire, N., & Taiet, G. (1970). *Can. J. Phys.* **48**, 1949
336 Teare, J.D., Taylor, R.L., & von Rosenberg, C.W. (1970). *Nature*, Lond. **225**, 240
337 Rhodes, C.K., Kelley, M.U., & Javan, A. (1968). *J. chem. Phys.* **48**, 5730
338 Pritchard, H.O., Sowden, R.G., & Trotman-Dickenson, A.F. (1953). *Proc. R. Soc.* **A217**, 563
339 Pritchard, H.O., Sowden, R.G., & Trotman-Dickenson, A.F. (1953). *Proc. R. Soc.* **A218**, 416
340 Richman, D.C., Lev-on, M., & Millikan, R.C. (1975). *Int. J. chem. Kinet.* **7**, 33
341 Stewart, J.L. & Stewart, E.S. (1952). *J. acoust. Soc. Amer.* **24**, 194
342 Winter, T.G. & Hill, G.L. (1967). *J. acoust. Soc. Amer.* **42**, 848
343 Sluijter, C.G., Knaap, H.F.P., & Beenakker, J.J.M. (1964). *Physica* **30**, 745
344 Jonkman, R.M., Prangsma, G.J., Ertas, I., Knaap, H.F.P., & Beenakker, J.J.M. (1968). *Physica* **38**, 441
345 Prangsma, C.J., Borsboom, L.J.M., Knaap, H.F.P., Van den Meidenberg, C.J.N., & Beenakker, J.J.M. (1972). *Physica* **61**, 527
346 Jonkman, R.M., Prangsma, G.J., Keijser, R.A.J., Aziz, R.A., and Beenakker, J.J.M. (1968). *Physica* **38**, 451
347 Prangsma, G.J., Heemskerk, H.F., Knaap, H.F.P., & Beenakker, J.J.M. (1970). *Physica* **50**, 433
348 Peterson, L.M., Lindquist, G.K., & Arnold, C.B. (1974). *J. chem. Phys.* **61**, 2480
349 Kneser, H.O., Bauer, H.-J., & Kosche, H. (1967). *J. acoust. Soc. Amer.* **41**, 1029
350 Breazeale, M.A. & Kneser, H.O. (1960). *J. acoust. Soc. Amer.* **32**, 885
351 Geide, K. (1963). *Acustica* **13**, 31
352 Holmes, R., Jones, G.R. & Pusat, N. (1964). *Trans. Faraday Soc.* **60**, 1220
353 Kelley, B.T. (1957). *J. acoust. Soc. Amer.* **29**, 1005
354 Holmes, R., Jones, G.R., Pusat, N., & Tempest, W. (1962). *Trans. Faraday Soc.* **58**, 2342
355 Greenspan, M. (1959). *J. acoust. Soc. Amer.* **31**, 155
356 Carnevale, E.H., Carey, C., & Larson, G. (1967). *J. chem. Phys.* **47**, 2829
357 Bauer, H. -J. & Kosche, H. (1966). *Acustica* **17**, 96
358 Kistemaker, P.G., Tom, A., & de Vries, A.E. (1970). *Physica* **48**, 415
359 Prangsma, G.J., Alberga, A.H., & Beenakker, J.J.M. (1973). *Physica* **64**, 278
360 Holmes, R., Jones, G.R., & Lawrence, R. (1964). *J. chem. Phys.* **41**, 2955
361 Kistemaker, P.G. & de Vries, A.E. (1975). *Chem. Phys.* **7**, 371
362 Valley, L.M. & Amme, R.C. (1969). *J. chem. Phys.* **50**, 3190
363 Kistemaker, P.G., Hanna, M.M., Tom, A., & de Vries, A.E. (1972). *Physica* **60**, 459
364 Kistemaker, P.G., Hanna, M.M., & de Vries, A.E. (1974). *Physica* **78**, 457
365 Widom, B. (1960). *J. chem. Phys.* **32**, 913
366 Wang Chang, C.S. & Uhlenbeck, G.E. (1951). *University of Michigan Report*, *CM-681*, Project N-Ord – 7294

367 Sather, N.F. & Dahler, J.S. (1961). *J. chem. Phys.* **35**, 2029
368 Sather, N.F. & Dahler, J.S. (1962). *J. chem. Phys.* **37**, 1947
369 Shields, F.D. (1970). *J. acoust. Soc. Amer.* **47**, 1262
370 Takayanagi, K. (1973). *Comments At. & Mol. Phys.* **4**, 59
371 Beckerle, J. (1953). *J. chem. Phys.* **21**, 2034
372 Brout, J. (1954). *J. chem. Phys.* **22**, 934
373 Takayanagi, K. (1957). *Proc. phys. Soc.* **A70**, 348
374 Takayanagi, K. (1959). *Scient. Rep. Saitama Univ.* **A3**, 87
375 Davison, W.D. (1962). *Disc. Faraday Soc.* **33**, 71
376 Davison, W.D. (1964). *Proc. R. Soc.* **A280**, 227
377 Agrawal, P.M. & Saksena, M.P. (1975). *Chem. Phys. Lett.* **32**, 123
378 Agrawal, P.M. & Saksena, M.P. (1974). *J. chem. Phys.* **61**, 848
379 Brout, J. (1954). *J. chem. Phys.* **22**, 1189
380 Agrawal, P.M. & Saksena, M.P. (1975). *J. Phys.* **B8**, 1575
381 Valley, L.M. & Amme, R.C. (1968). *J. acoust. Soc. Amer.* **44**, 1144
382 Takayanagi, K. (1971). *Inst. Space and Aeronaut. Sci. (Tokyo)* **36**, 229 (Report No. 467)
383 Zeleznik, F.J. & Dugan, J.V. (1971). *J. chem. Phys.* **54**, 4523
384 Saksena, M.P., Agrawal, P.M. & Harminder (1973). *Ind. J. Pure & Appl. Phys.* **11**, 563
385 Broida, H.P. & Carrington, T. (1963). *J. chem. Phys.* **38**, 136
386 Steinfeld, J.I. & Klemperer, W. (1965). *J. chem. Phys.* **42**, 3475
387 Rhodes, J.E. (1946). *Phys. Rev.* **70**, 932
388 Brau, C.A. & Jonkman, R.M. (1970). *J. chem. Phys.* **52**, 477
389 Narayana, H. & Woods, S.B. (1970). *Can. J. Phys.* **48**, 305
390 Crawford, O.H. (1970). *Chem. Phys. Lett.* **6**, 409
391 Rabitz, H. & Lam, S.-H. (1975). *J. chem. Phys.* **63**, 3532
392 Ding, A.M.G. & Polanyi, J.C. (1975). *Chem. Phys.* **10**, 39
393 Unland, M.L. & Flygare, W.H. (1966). *J. chem. Phys.* **45**, 2421
394 Gordon, R.G. (1967). *J. chem. Phys.* **46**, 4309

SUBJECT INDEX

CHEMICAL INDEX

This index is based on the procedure followed in the index to ref. 89. All individual molecules (chemical compounds) are listed under their chemical formulae, as written in the text, considered as words; for example, $CHCl_3$ under Chcl, CH_3OCH_3 under Choch. If there are several molecules giving the same 'word' they are listed in the order of increasing numbers of the first, second, third, . . . atom; for example $CHCl_3$, CH_2Cl_2, CH_3Cl, $C_2H_2Cl_4$, $C_2H_4Cl_2$ in this order, but C_6H_{14} is ahead of $CHCl_3$ since the corresponding 'word' is Ch. Prefixes such as c-, n-, *o*-, *p*-, are disregarded, except that *o*-H_2 follows H_2 and precedes *p*-H_2.

Page numbers in bold type refer to information about energy relaxation of the molecule itself.